Springer Tracts in Advanced Robotics 115

Editors

More information about this series at http://www.springer.com/series/5208

Carlos A. Cifuentes · Anselmo Frizera

Human-Robot Interaction Strategies for Walker-Assisted Locomotion

Springer

Carlos A. Cifuentes
Department of Biomedical Engineering
Colombian School of Engineering Julio Garavito
Bogotá
Colombia

Anselmo Frizera
Department of Electrical Engineering
Federal University of Espirito Santo
Goiabeiras Vitória
Brazil

ISSN 1610-7438 ISSN 1610-742X (electronic)
Springer Tracts in Advanced Robotics
ISBN 978-3-319-81669-2 ISBN 978-3-319-34063-0 (eBook)
DOI 10.1007/978-3-319-34063-0

Printed on acid-free paper

This Springer imprint is published by Springer Nature
The registered company is Springer International Publishing AG Switzerland

I have not failed. I've just found 10,000 ways that won't work.

—Thomas A. Edison

It is not true that people stop pursuing dreams because they grow old, they grow old because they stop pursuing dreams.

—Gabriel García Márquez

Series Foreword

Robotics is undergoing a major transformation in scope and dimension. From a largely dominant industrial focus, robotics is rapidly expanding into human environments and vigorously engaged in its new challenges. Interacting with, assisting, serving, and exploring with humans, the emerging robots will increasingly touch people and their lives.

Beyond its impact on physical robots, the body of knowledge robotics has produced is revealing a much wider range of applications reaching across diverse research areas and scientific disciplines, such as biomechanics, haptics, neurosciences, virtual simulation, animation, surgery, and sensor networks. In return, the challenges of the new emerging areas are proving an abundant source of stimulation and insights for the field of robotics. It is indeed at the intersection of disciplines that the most striking advances happen.

The *Springer Tracts in Advanced Robotics* (STAR) is devoted to bringing to the research community the latest advances in the robotics field on the basis of their significance and quality. Through a wide and timely dissemination of critical research developments in robotics, our objective with this series is to promote more exchanges and collaborations among the researchers in the community and contribute to further advancements in this rapidly growing field.

The monograph by Carlos A. Cifuentes and Anselmo Frizera presents the outcome of recent research results in the field of rehabilitation robotics. A number of human-robot interaction (HRI) strategies are introduced for walker-assisted locomotion. Remarkably, both physical and cognitive HRI concepts are considered for the development of assistive technologies. A new multimodal human-robot interface for testing and validating control strategies applied to robotic walkers for assisting human mobility and gait rehabilitation is presented. Trends and opportunities for future advances in the field of assistive locomotion via the development of hybrid solutions based on the combination of smart walkers and biomechatronic exoskeletons are also discussed.

Most methods have been effectively implemented in experimental tests under real rehabilitation conditions. A fine addition to STAR!

Naples, Italy
March 2016

Bruno Siciliano
STAR Editor

Preface

Neurological and age-related diseases affect human mobility at different levels causing partial or total loss of such faculty. There is a significant need to improve safe and efficient ambulation of patients with gait impairments. In this context, walkers present important benefits for human mobility, improving balance and reducing the load on lower limbs. Most importantly, walkers induce the use of patient's residual mobility capacities in different environments. In the field of robotic technologies for gait assistance, a new category of walkers has emerged, integrating robotic technology, electronics, and mechanics. Such devices are known as "robotic walkers," "intelligent walker," or "smart walkers."

One of the specific and important common aspects to the field of assistive technologies and rehabilitation robotics is the intrinsic interaction between the human and the robot. In this book, the concept of human-robot interaction (HRI) for human locomotion assistance is explored. This interaction is composed of two interdependent components. On the one hand, the key role of a robot in a physical HRI (pHRI) is the generation of supplementary forces to empower human locomotion. This involves a net flux of power between both actors. On the other hand, one of the crucial roles of a cognitive HRI (cHRI) is to make the human aware of the possibilities of the robot while allowing him to maintain control of the robot at all times.

This book will also present a new multimodal human-robot interface for testing and validating control strategies applied to robotic walkers for assisting human mobility and gait rehabilitation. This interface extracts navigation intentions from a novel sensor fusion method that combines: (i) a laser range finder (LRF) sensor to estimate the users legs' kinematics, (ii) wearable inertial measurement Units (IMUs) to capture human and robot orientations, and (iii) two triaxial force sensors measure the physical interaction between the human's upper limbs and the robotic walker.

Two close control loops were developed to naturally adapt the walker position and to perform body-weight support strategies. First, a force interaction controller generates velocity outputs to the walker based on the upper-limbs physical interaction. Second, a inverse kinematic controller keeps the walker within a desired position to the human improving such interaction.

The proposed control strategies are suitable for natural human-robot interaction as shown during the experimental validation. Moreover, methods for sensor fusion for estimating control inputs were presented and validated. In the experimental studies, the parameters estimation was precise and unbiased. It also showed repeatability when speed changes and continuous turns were performed.

At the end, this book will focus on describing the upcoming research in the field of assisted locomotion which leads to the development of hybrid solutions based on the combination of smart walkers and biomechatronic exoskeletons. Additionally, new technological breakthroughs regarding human-robot interaction taking into account the environment are also defined. In this manner, the aim is to achieve a closer interaction between the robotic solution and the individual empowering the rehabilitation potential of such devices in clinical applications.

Vitoria-ES, Brazil
March 2016

Carlos A. Cifuentes
Anselmo Frizera

Acknowledgements

The work presented in this book has been carried out with the financial support from the National Council of Technological and Scientific Development (CNPq), Espirito Santo Research and Innovation Foundation (FAPES), and Coordination for the Improvement of Higher Education Personnel (CAPES) from Brazil.

The future work defined in the book is ongoing by means of a cooperation research among different groups: the Robotics and Industrial Automation Group at Federal University of Espirito Santo (Brazil), the Biomedical Engineering Group at Colombian School of Engineering Julio Garavito (Colombia), the Neural and Cognitive Engineering Group at Spanish Council for Scientific Research (Spain), and the Automation Institute at National University of San Juan (Argentina).

We wish to express our gratitude to Prof. Teodiano Freire Bastos Filho, Prof. Ricardo Carelli, and Prof. Eduardo Rocon, who were willing to cooperate in different ways in this research, and also to take the job of reading and evaluating this work.

Furthermore, we would like to thank several students who helped us in developing and implementing some ideas of this work. In particular, we are very grateful to Camilo Arturo Rodríguez Díaz, who did substantial work on the development of the *UFES Smart Walker*.

Contents

Abbreviations

ADL	Activities daily living
BWS	Body-weight support
C	Center of rotation
cHRi	cognitive human-robot interface
cHRI	cognitive human-robot interaction
CP	Cerebral palsy
FLC	Fourier linear combiner
FQI	Filter quality indicator
GC	Gait cadence
H	Human
HAT	Head arms trunk
HCI	Human-computer interaction
HCi	Humancomputer interface
HIZ	Human interaction zone
HMI	Human-machine interaction
HMi	Human-machine interface
HRI	Human-robot interaction
IMU	Inertial measurement init
LDD	Legs Difference Distance
LL	Left leg
LMS	Least mean square
LRF	Laser range finder
MEMS	Microelectromechanical system
pHRi	physical human-robot interface
pHRI	physical human-robot interaction
R	Robot
RL	Right leg
RMSE	Root mean square error

SCI	Spinal cord injury
SL	Step length
W	Walker
WFLC	Weighted-frequency FLC

Symbols

Symbols	Name	Unit
v_h	Human linear velocity	m/s
ω_h	Human angular velocity	°/s
ψ_h	Human orientation (pelvic orientation)	°
v_r	Robot linear velocity	m/s
ω_r	Robot angular velocity	°/s
ψ_r	Robot orientation	°
v_w	Walker linear velocity	m/s
ω_w	Walker angular velocity	°/s
ψ_w	Walker orientation	°
$\overline{RH}$	Human-robot line	m
$\overline{RC}$	Robot axis line	m
$\overline{WH}$	Human-walker line	m
$\overline{WC}$	Walker axis line	m
φ	Angle between v_h and $\overline{RH}/\overline{WH}$	°
θ	Angle between $\overline{RH}/\overline{WH}$ and $\overline{RC}/\overline{WC}$	°
d	Human-robot distance	m
dd/d_d	Desired human-robot distance	m
$\tilde{d}$	Distance error	m
$\tilde{\varphi}$	φ error	°
k_d	Constant to adjust $\tilde{d}$	
k_φ	Constant to adjust $\tilde{\varphi}$	
d_1/LL	Left leg's distance	m
d_2/RL	Right leg's distance	m
θ_1	Left leg's angle	°
θ_2	Right leg's angle	°
$v_r(C)$	Control action v_r	m/s

(continued)

(continued)

Symbols	Name	Unit
$\omega_r(C)$	Control action ω_r	°/s
$v_r(R)$	Measured v_r	m/s
$\omega_r(R)$	Measured ω_r	°/s
$v_w(C)$	Control action v_w	m/s
$\omega_w(C)$	Control action ω_w	°/s
$v_w(R)$	Measured v_w	m/s
$\omega_w(R)$	Measured ω_w	°/s
K	Constant ratio related to LRF measured and step length	
M	Number of harmonics of the Fourier model	
x_k	FLC reference signal	
s_k	FLC oscillatory component	
$\hat{s}_k$	FLC estimated oscillatory component	
ε_k	$y_k - \hat{s}_k$	
v_k	FLC stationary input component	
y_k	FLC input signal $s_k + v_k$	
$\mathbf{W}_k$	FLC Fourier series coefficients	
μ	FLC amplitude adaptation gain	
ω_{0_k}	WFLC adaptive frequency	
$\omega_{0,0}$	WFLC instantaneous frequency	
μ_0	WFLC-updated frequency weight	
μ_1	WFLC-updated amplitude weight	
μ_b	WFLC bias weight	
$\|\mathbf{W}_k\|$	Magnitude of the WFLC $\mathbf{W}_k$	
F_{ry}	Right arm y-axis forces	Kgf
F_{ly}	Left arm y-axis forces	Kgf
F_{rz}	Right arm z-axis forces	Kgf
F_{lz}	Left arm z-axis forces	Kgf
F_r	F_{ry}/F_{rz}	
F_l	F_{ly}/F_{lz}	

List of Figures

List of Tables

Chapter 1
Assistive Devices for Human Mobility and Gait Rehabilitation

Mobility is one of the most important human faculties and can be defined as the ability of an individual to move freely through multiple environments and perform daily personal tasks with ease [1]. Neurological and age-related diseases affect human mobility at different levels causing partial or total loss of such faculty. In addition, mobility decreases gradually with age. Evidences show that mobility restrictions are also associated with cognitive and psychosocial disturbances, which further impairs the quality of life of the individual [2]. In this context, new technologies have emerged to improve the life conditions of people with motor impairments. Some remarks regarding the human locomotor system, mobility dysfunctions, assistive devices for enhancing mobility, functional compensation during walking and devices for gait rehabilitation will be presented in this chapter.

1.1 The Locomotor System

The locomotor system is composed of bones with their surrounding tissues, such as cartilage, muscles, ligaments, the nervous system controlling the motorics, and connective tissue. The locomotor function is studied by subsystems [3]: the skeletal subsystem supports the body as a whole, the skeletal muscle plant enables active force generators with the purpose of achieving movement. That way, active muscular forces combined with external forces (gravitational, ground reaction), elastic muscle forces, and other inertial forces which occur due to the moving body mass, all determine manifested body kinematics in time and space. Finally, the control subsystem performed by the nervous system, which coordinates and controls overall motor activity.

As active force generators in the realization of movement, skeletal muscles are coordinated in time and space and through contraction intensities when a particular movement structure is performed. According to [4]: "The course of a movement is nothing else but a projection to the outside of a pattern of excitation taking place

C.A. Cifuentes and A. Frizera, *Human-Robot Interaction Strategies for Walker-Assisted Locomotion*, Springer Tracts in Advanced Robotics 115,
DOI 10.1007/978-3-319-34063-0_1

in a corresponding setting in the central nervous system". In this way, stereotypical locomotor activities, such as walking, manifest, to a great extent, the automatism of excitation and inhibition patterns and, in concordance with this, the defined order and contraction intensity of the corresponding musculature.

Central Pattern Generators (CPGs) are neural circuits located in the end parts of the brain and first parts of the spinal cord of a large number of animals and are responsible for generating rhythmic and periodic patterns of locomotion in different parts of their bodies [5]. Although these pattern generators use very simple sensory inputs imported from the sensory systems, they can produce high dimensional and complex patterns for walking, swimming, jumping, turning and other types of locomotion [6].

In disturbed function of the locomotor apparatus due to disease or injury, the natural automatism of stereotypical locomotor activities is disrupted. Rehabilitation, then, aims at restoring the disturbed function or, in the case of permanent anatomical or functional changes or deficiencies, modifying the function of the neuro-musculo-skeletal system so that the role of the insufficient or lost muscles is taken over by the remaining (healthy) musculature [7].

1.2 Conditions that Affect Mobility

Different conditions, such as stroke, spinal cord injury and cerebral palsy affect human mobility causing partial or total loss of locomotion capacities. In addition, it is known that mobility decreases gradually with age as a consequence of neurological, muscular and/or osteoarticular deterioration. The next paragraphs will present general definitions and statistics regarding such conditions.

A stroke is the consequence of cell death within the brain relating to either internal bleeding or a blockage in one of the two main supplying arteries. Currently, it represents a major problem in clinical medicine being a leading cause of disability in the developed world [8]. Neurological impairments after stroke have tendencies of causing hemiparesis or partial paralysis, which can deprive patients of performing activities of daily living (ADL) like walking.

Stroke survivors typically show significantly reduced gait speed, shortened step length and loss of balance in their gait patterns and often experience falls [9]. With the proven fact that repetitive and persistent stimulation could restore and reorganize defective motor functions caused by neurological disorders, there is a strong need for new therapeutic interventions [10].

Spinal Cord Injury (SCI) consists of any alteration of the spinal cord that affects the sensory-motor and autonomous systems under the level of lesion. Based on a conservative average of annual incidence of 22 people/million population in the western and developing world, it is estimated that over 130,000 people each year survive a traumatic spinal cord injury and begin a "new and different life" bound to a wheelchair for 40 years or more [11].

SCI is a devastating clinical circumstance due to the functional loss resulting on a great impact on the functional independence of the person, affecting the quality of life, life expectancy and causing important economic problems, considering the costs associated with primary care and loss of income. Rehabilitation of SCI is aimed towards the maximization of user independence and adequate management of secondary lesion-related diseases. Maximization of mobility has been identified as one of the main objectives for the injured individuals [12].

Cerebral palsy (CP) is a disorder of posture and movement due to a defect or lesion in the immature brain [13]. The impairments are permanent, but not unchanging, and cause activity limitation and participation restrictions. CP is the most common cause of permanent serious physical disability in childhood, with an overall prevalence of around 2 per 1000 live births. It is estimated that 650,000 families in Europe either have a child with CP or support an adult with CP [14].

The prospect of survival in children with severe level of impairment has increased in recent years. That way, new strategies are needed to help to promote, maintain, and rehabilitate the functional capacity, and thereby diminish the dedication, the required assistance and the economical demands that this condition represents for the patient, the caregivers and the society [15].

Finally, it is important to mention that the world's population over 60 years old will more than triple (from 600 million to 2 billion) between the years 2000 and 2050 [16]. The majority of this increase is occurring in less developed countries where this group will rise from 400 million, in the year 2000, to 1.7 billion by the year 2050 [16].

Disordered gait, defined as a gait that is slowed, aesthetically abnormal, or both, is not necessarily an inevitable consequence of aging, but rather a reflection of the increased prevalence and severity of associated diseases [17]. Common diagnosis among people over 60 years old also include cardiovascular conditions, dementia, diabetes, arthritis, osteoporosis, and stroke [18]. These conditions all have the potential to impact the human mobility. Thus, there is a significant need to improve the ability for older adults to have safe and efficient ambulation, as this may help to reduce the incident of fall and fractures. That way, some studies have shown that walking programs with a frequency of at least 3–5 times per week have been found to increase walking endurance and distance [19].

1.3 Emerging Design Approaches for Assistive Devices

The machines have changed during the modern times as consequence of the incorporation of electronics, allowing the integration of sensors and control, and the evolution of human-machine interfaces. This gave origin to mechatronics as the modern paradigm of machine design and the baseline for the development of robotics [20].

The new complete scheme can be seen in an analogy with biological systems, integrating a musculoskeletal apparatus with a nervous system and a circulatory apparatus. When machine design takes inspiration from biology, as in this analogy, then it can be referred to as biomechatronics. Robotics, especially, is now following this direction, with a stronger emphasis on biorobotics and biomedical applications [20].

Biomechatronic systems integrate mechanisms, sensors, control strategies, human-robot interaction and signal processing techniques. These components are inspired by biological models. This book stress the biomechatronic conception of robotic walkers.

According to Pons [21], the scope of biomechatronics is broader in three distinctive aspects: firstly, biomechatronics intrinsically includes bioinspiration in the development of mechatronic systems, e.g. the development of bioinspired mechatronic components (control architectures, actuators, etc.); secondly, biomechatronics deals with mechatronic systems in close interaction with biological systems, e.g. wearable robot interacting with a human; and, finally, biomechatronics commonly adopts biologically inspired design and optimization procedures in the development of mechatronic systems, e.g. the adoption of adaptable algorithms in the optimization of mechatronics components.

Another design approach is the neuro-robotics paradigm, mainly aimed at the fusion of neuroscience and robotic competences and methods to design better robots that can act and interact closely with humans, the neuro-robotics paradigm applies in the research area of 'human augmentation' through 'hybrid bionic systems'. According to E. Von Gierke, a pioneer of this discipline, the primary goal of bionics is "to extend man's physical and intellectual capabilities by prosthetic devices in the most general sense, and to replace man by automata and intelligent machines" [22].

Hybrid Bionic Systems (HBSs) can be generically defined as systems that contain both technical (artificial) and biological components. In recent years, many scientific and technological efforts have been devoted to create HBSs that link, via neural interfaces, the human nervous system with electronic and/or robotic artefacts. In general, this research has been carried out with various aims: on the one hand, to develop systems for restoring motor and sensory functionalities in injured and disabled people; on the other hand, for exploring the possibility of augmenting sensory-motor capabilities of humans in general, not only of disabled people [23].

In future, electroencephalography (EEG) signals from the brain, or electromyography (EMG) signals from body muscles, may be converted into electronic commands that will allow control of the robots. A working robot powered by the wearer's thoughts is possible by using a brain-computer interface (BCI), which is a hybrid system that records brain signals and decodes the user's intention to operate and to control advanced devices, while also providing feedback to the user about brain activity.

1.4 Mobility Assistive Devices

Although most gait/mobility disturbances are well recognized, only a fraction of such conditions can be fully reversed by surgical procedures or rehabilitation approaches. Therapeutic alternatives, in such cases, include the selection and prescription of assistive devices to provide adequate functional compensation and to stop the progression of the disability and to improve the overall quality of life of the affected subjects [24].

The choice of the most appropriate assistive device requires careful analysis and interpretation of the clinical features associated with the subject's residual motor capacities, including cognitive function, vision, vestibular function, muscle force (trunk and limbs), degenerative status of lower and upper limb joints, overall physical conditioning of the patient and also additional characteristics of the environment in which the patient lives and interacts. Severe dysfunctions in one or more of such features can compromise the safe use of the device and increase the risk of falls or compromise locomotion performance due to energy expenditure [25].

Based on the levels of mobility restriction, the patients may be classified into two broad functional groups: (i) individuals with total loss of the mobility capacity and (ii) individuals with partial loss of mobility, presenting different levels of residual motor capacity.

Individuals belonging to the first group may have completely lost the ability of move by themselves and are at high risk of confinement in bed and, consequently, to suffer the effects of prolonged immobility. Examples of subjects in this functional group include patients with complete spinal cord injury, advanced neurodegenerative pathologies, severe lower limb osteoarthritis and fractures of the spine/lower limb bones. In such cases, however, the motion can be performed by assistive technology known as the *alternative devices*. Without the use of such technologies, the locomotion may become an impossible task for these individuals [26]. Some examples of alternative devices are (robotic) wheelchairs and special vehicles, including adapted scooters.

The mobility provided by alternative devices can help patients to gain a certain amount of independence during daily tasks and may have positive impact on self-esteem and social interaction. However, the prolonged use of such devices do not prevent immobility-related adaptations in spine and lower limbs, characterized by loss of bone mass, circulatory disorders, pressure ulcers and other physiological impairments [27].

The second functional group is composed of individuals that present some level of residual motor capacity, which can be empowered by an assistive device. In other words, the use of such *augmentative devices* aims to empower the user's natural means of locomotion, taking advantage of the remaining motor capabilities. In the last decade, researches in the field of intelligent augmentative devices have increased, with focus on the implementation of advanced robotic solutions for people with disability and robot-assisted rehabilitation therapy interventions for motor recovery after neurologic injuries. The augmentative devices can be further classified into: (i) self-ported or wearable and (ii) external devices.

The wearable devices are mainly represented by orthoses and protheses. An orthosis is a mechanical structure that maps on to the anatomy of the human limb. Its purpose is to restore lost or weak functions. A prosthesis is an electromechanical device that substitutes lost limbs after amputation.

External devices are represented by canes, crutches and walkers. Considering that such devices are focus of this book, they will be addressed with more detail in the next section.

1.5 Devices for Functional Compensation of Gait

Human gait starts as a nerve impulse in the central nervous system and ends with the generation of the ground reaction forces [28]. Conventionally, the heel strike is used for dividing the gait cycles. The gait cycle is divided into two phases: stance and swing. Both the beginning and the end of stance involve a period of bilateral foot contact with the floor (double support). Alternatively, during the swing phase, the foot is in the air and the leg is swinging through preparation for the next foot strike [29].

Patients with mobility dysfunctions present significant functional limitations, including the inability to support the body weight through the lower-limbs, to generate propulsive forces, to move the limbs swiftly through an appropriately timed trajectory and to control lateral stability. They employ compensatory strategies to continue forward propulsion with a stable base of support. Internally based compensatory strategies include reduced gait velocity, increased stance and double support time, knee hyperextension in stance, and hip circumduction during swing phase [30]. External devices provide weight support during walking and enable functional compensation strategies to improve the patient's mobility. Benefits and limitations of theses devices are discussed as follows.

Canes are more commonly used to increase gait stability rather than to partial weight-support. A simple single point cane may prevent or reduce falls in patients with imbalance. Crutches allow a direct support of the body, thus providing great stability and balance in walking and a greater weight support compared with canes. However, crutches are cumbersome and their use may lead to unnatural gait patterns. Therefore, the amount of weight support is neither constant nor quantifiable because it depends on the strength of the patient as well as the degree of control of their upper-limbs and trunk [31].

Other external devices are the walkers, which are characterized for their structural simplicity, low cost and great rehabilitation potential. Walkers are usually prescribed for patients in need of gait assistance, to increase static and dynamic stability and also to provide partial body weight support during functional tasks [32]. Such devices empower the residual motor capacities of the user, allowing a natural way of locomotion and, thus, preventing immobility-related changes. Additionally, evidence shows that walker-assisted gait is related to important psychological benefits, including increased confidence and safety perception during ambulation [33].

The *standard frame* (Fig. 1.1a) is the most common configuration of a passive walker. It is based on a metal frame with four rigid legs that must contact the ground simultaneously during each step. It is considered the most stable model, but requires a slow and controlled gait pattern, since the user must lift the device completely off the ground and move it ahead before taking a step forward [25].

Critics regarding the use of standard frames arise from evidence of increased force levels exerted by the upper limbs during locomotion [34]. The gait pattern imposed by the device also increases the user's energy expenditure by 217 % during level walking when compared to unassisted or wheeled walker-assisted gait (presented

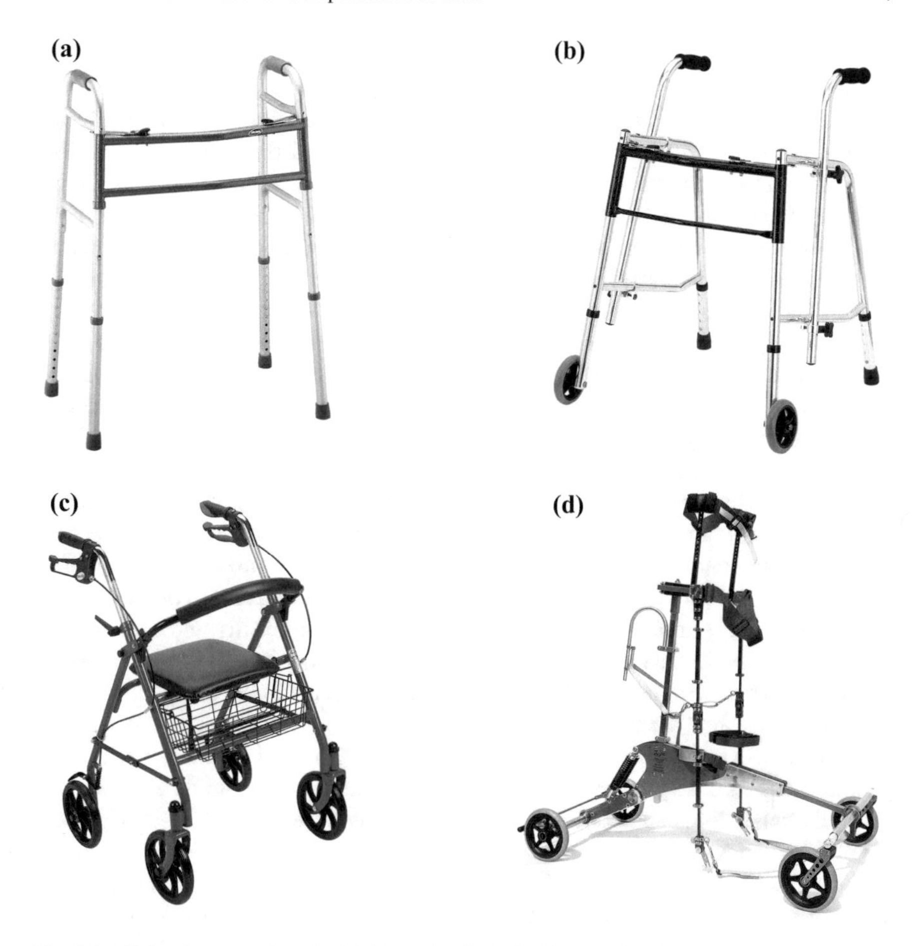

Fig. 1.1 Walker frames. **a** Standard. **b** Two-wheeled. **c** Rollator. **d** Hands-free (NF-Walker—Made for Movement)

below). Such findings restrict the prescription of standard walkers for patients presenting severe levels of metabolic, cardiac or respiratory dysfunctions [35]. Patients with cognitive disorders are also not among the scope of potential users of standard frames. This recommendation is mainly based on the results presented in [36], which reported that gait assisted by standard walkers requires higher levels of attention when compared to canes or other walker models to avoid the risk of falls.

The *two-wheeled walkers* (Fig. 1.1b) are another variation of conventional walkers. Although similar to standard frames in many aspects, these versions are characterized by the presence of two wheels mounted on the front legs (front-wheeled walkers). Such models are recommended for more active subjects or patients that have a hard time in lifting the device from the ground. The wheels allow the performance of a more natural gait pattern, but evidence shows that dynamic stability during walking is lower than standard frame assistance, and the energy expenditure is 84 % higher when compared to normal ambulation [24, 25].

Rollator walkers (Fig. 1.1c) can be seen as a modification of the two-wheeled models. They present four wheels attached to the legs of the walker. These models allow faster locomotion and better performance of a natural gait pattern during locomotion and also present lower energetic expenditure compared to other walker models. However, rollators are considered the most unstable walker version and the risk of falls is significantly increased in situations that require full body-weight support of the user, due to uncontrolled displacement of the device. In a clinical setting, rollators may be recommended for patients that require a broad walking base without the need of continuous body-weight support. The design of such models allows great number of adaptations, like the installation of breaking system at the handles (to increase static stability), different wheel sizes, robust frames, adaptation of seat cushions, among others [24, 35].

Another assistive device is the *hands-free walker* (Fig. 1.1d). It includes adaptations in order to minimize weight bearing by the upper extremities at the same time as it promotes continuous body-weight support. Hands-free walkers mutually differ in where they connect to the subject, in how many degrees of freedom of motion they allow, and in how driven/steerable/actuated they are. Complexity and cost are also points to observe. Some of the devices offer the possibility to assist in sit-to-stance transition, to perform turns during walking or to be automatically driven and steered [37]. These walkers assure the safety of a person that is walking, which is a strong need for individuals with hemiplegia, as most of them need assistance with their posture while standing, as well as with the swing phase of the paretic leg [38]. It is also common to find hands-free walkers integrated with wearable orthoses in devices for children with CP [39].

In the field of robotic technologies for gait assistance, there are several ongoing projects regarding robotic versions of walkers and other guidance devices. In this context, a new category of walkers has emerged, integrating robotic technology, electronics and mechanics. Such devices are known as "robotic walkers", "intelligent walkers" or "smart walkers" [40]. Smart walkers present a great number of functionalities and are capable of providing mobility assistance at different functional levels, better adjusted to the individual needs of the user [40, 41].

Robotic walkers are usually mounted over a rollator frame. This configuration takes advantage of the versatility of the four wheels and the ability to maintain approximate natural patterns of walking. Stability issues are approached with special security mechanisms to prevent falls and undesirable movement intentions from the user [41]. The development of Human Machine Interfaces (HMI) to interpret the user's commands enables the implementation of different control strategies, which may allow safer human-walker mobility. Consequently, HMI in smart walkers will be broadly discussed in the next chapter.

1.6 Trends in Gait Rehabilitation

Although the majority of patients with mobility dysfunctions using augmentative devices achieve some level of ambulation, there continues to be a strong need for therapeutic interventions that can reduce the long-term need for physical assistance and result in a biomechanically efficient and stable gait pattern that does not diminishes over time. However, the conventional rehabilitation procedures require excessive laboring efforts of therapists in assisting walking of severely affected subjects, setting the paretic limb and controlling trunk movements. Under those circumstances, rehabilitation process is often not completely successful for reasons, such as: limited amount of walking space in each therapy session, exhaustion, falling, injuries and the patient's fear of falling.

In this context, Body Weight Support (BWS) Systems have come to play an important role in gait rehabilitation. Partial unloading of the body weight allows neurologically challenged patients with weak muscles to practice gait training more efficiently. Furthermore, robot-assisted rehabilitation therapy is as an emerging form of rehabilitation treatment for motor recovery after neurological injuries. Robotic devices can help patients achieve the intensive, repetitive practice needed to stimulate neural recovery, reducing the need for supervision and improving cost-benefit profiles [42].

Treadmill based devices are the most prevalent robotic rehabilitation methods, and Lokomat (Hocoma, Switzerland) [43] is the most clinically tested system and one of the firsts of its type. In such device, pelvic vertical movements, hip and knee joints are driven by orthoses linked to the treadmill frame. Other related devices are also available: AutoAmbulator (HealthSouth, US) [44], Lopes [45], ALEX [46], PAM & POGO [47]. However, this type of systems provides only forward movement with predetermined paths to ensure accurate reproduction of physiological kinematics. This way, patients are constrained to a fixed platform and a predetermined gait pattern, which is not natural and leads to less satisfactory functional outcomes. In contrast, a growing body of clinical studies suggests that effective training in neurorehabilitation allows subjects to participate actively and perform unhindered movements, according to strategies like "assist-as-needed" [48] or the "challenge point" concept [49].

Additionally, there are controversy on the assumption that walking on a treadmill could represent an actual gait on the over-ground in terms of body mass shifting, body mass acceleration and sensorimotor feedback (such as proprioceptive inputs). Walking on treadmill indicates significantly greater cadence, smaller stride length and stride time as well as reductions in the majority of joint angles, moments, powers and pelvic rotation excursion compared with over-ground walking [50, 51].

Finally, as a matter of fact, over-ground walking is considered as the most natural gait pattern with actual foot contact. Thus, over-ground walking rehabilitation devices are recommended for increasing gait performance as well as having natural gait patterns. In this context, mobile gait rehabilitation devices that combine mobile platforms (smart walkers) with BWS system can enable free walking over ground in different environments such as outside or at home, offering more realistic, flexible and motivating training conditions.

1.7 Background

This work is based on previous researches which were developed either inside of our group or within the execution of collaboration projects.

The ASAS Project (translation from Spanish "Pseudo-robotic walker for enhancing user's security") introduced the development of a basic smart walker physical structure [26]. In this project, some important features were implemented on a commercial walker as it can be seen in Fig. 1.2a. The mechanical structure was modified to provide forearm supporting platforms, which are more comfortable and stable than the conventional handlebars. Such platforms also stabilize the trunk and the upper-limbs during the walker-assisted gait. Two motorized wheels were also installed allowing more control of the device's motion. It also included a basic HMI to control the walker's motion. Such interface is composed of two push-buttons located on each handlebar, at the height of the user's thumbs.

However, such approach presents some disadvantages regarding the user's interfaces. Each push-button commands a corresponding motorized wheel: when the right push-button is pressed, the right motor moves forward. That way, when the user presses both buttons, the walker moves forward. The user needed previous training and good motor coordination to properly command the robotic walker.

SIMBIOSIS, an improvement of a ASAS walker, was a sequential project that aimed to develop a multimodal interface that combines upper-limbs reaction forces and lower-limbs cadence estimation using ultrasonic sensors [52]. Such interface allowed natural interaction as the user's guiding intentions were detected and transmitted to the walker's control system. The SIMBIOSIS Walker is shown in Fig. 1.2b. Such project presents a complete study of different filtering strategies to extract

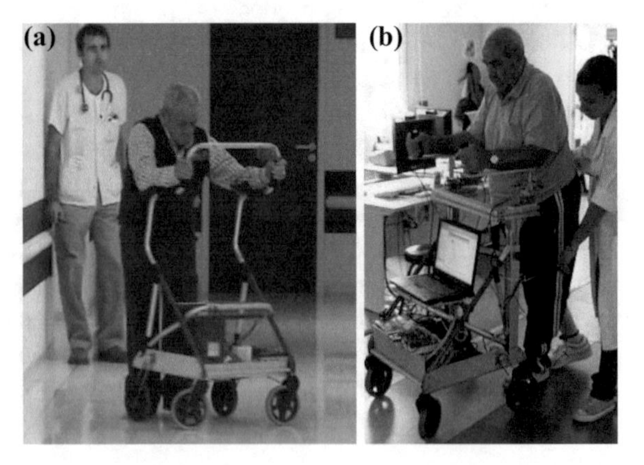

Fig. 1.2 Previous Smart Walkers related to this work. **a** ASAS Walker. **b** SIMBIOSIS Walker

guiding intentions from the upper-limb reaction forces acquired by force sensors installed in the forearm supporting platforms. As a conclusion of this study, a method to estimate the upper-limb motion intentions was developed. This work relies on such estimation algorithms for the extraction of the control inputs as presented in Sect. 4.3.2.

Nevertheless, the SIMBIOSIS Project proposed a basic control strategy only to demonstrate the effectiveness of the estimation algorithms. Although experimental results showed that this approach allows natural interaction with the device, robust control for human-walker interaction and the user's dependability were not taken into account. Therefore, the work presented in this book is focused on the development of natural and robust human-walker control strategies, which are based on the development of a new multimodal interface.

1.8 Scope of the Book

This book focuses on a human-robot interaction (HRI) strategy for human mobility assistance. The integration of HRI concepts in the smart walkers field is addressed to enable natural channels of communication between the walker and the human. Additionally, this book presents a multimodal human-robot interface that provides a means of testing and validating control strategies for robotic walkers for assisting human mobility and gait rehabilitation.

The primary objective of this work is to design new control strategies to develop a more natural, safer and more adaptable human-walker interaction. Beyond this general goal, there are several specific objectives that are presented below.

1. To study the human motion intentions during walker-assisted gait in order to extract human-walker interaction parameters.
2. To integrate HRI sensing modalities that promote natural human-walker interaction.
3. To design a multimodal interface for testing and validating control strategies for robotic walkers.
4. To design and validate a cognitive HRI control strategy for walker-assisted gait.
5. To develop a control strategy based on cognitive and physical HRI for walker-assisted gait.

The work presented in the following chapters will describe how such objectives are achieved. Moreover, the key contributions of this work are the development of a novel multimodal HRI control strategy and two new sensor fusion methods to estimate the control inputs. Objectively, the most important technical and scientific contributions of the research presented in this work are listed below.

1. Formulation of a control strategy for cognitive HRI during walking. Such controller was evaluated using a mobile robot and a robotic walker.

2. Proposal and validation of a new method to obtain human-robot interaction parameters synchronized with gait cycles and acquired with Laser Range Finder (LRF) and Inertial Measurement Units (IMU).
3. Proposal and validation of a new method to continuously estimate human-walker interaction parameters based on the adaptive estimation and filtering of gait components from LRF and IMU sensors.
4. Design and validation of a multimodal interface for walker-assisted gait based on the combination of LRF, IMU and 3D force sensors.
5. Development of a control strategy based on cognitive and physical HRI for walker-assisted gait.

This book is organized in five thematic chapters (Chaps. 2–6), addressing relevant aspects regarding walker assisted locomotion and the important interaction aspects involved in this process.

Chapter 2 addresses the literature review concerning human-robot interaction, paying a special attention to the interfaces that have been implemented or can be useful for human-walker interaction. That chapter also presents the concept of dual human-robot interaction in assisted locomotion.

Chapter 3 begins with the formulation of cognitive HRI control strategy for human tracking applied to a mobile robot. Afterwards, a method for estimation of control inputs is proposed and validated. The LRF and IMU sensors are introduced as sensor interfaces for human tracking. Finally, an experimental study is performed to validate both the estimation of interaction parameters and the control implementation using a mobile robot.

Chapter 4 addresses the integration of the control strategy proposed in Chap. 3 on a robotic walker. Some remarks regarding the human-robot physical link demanded a new human-walker parameters detection method, which was formulated and validated. That chapter also presents a new robotic walker platform to fulfill the sensor and interaction requirements. Finally, an experimental study is performed to validate both the control parameters detection and the control implementation.

Chapter 5 introduces upper-limb reaction forces as a physical HRI interface. A multimodal interface for human mobility assistance is presented. That interface is evaluated as a tool for understanding the human motion intentions during walker-assisted gait. A final control strategy based on physical and cognitive HRI is presented and validated to conclude the scope of this book.

Finally, Chap. 6 presents the conclusions and some recommendations for future works in this challenging field of rehabilitation robotics.

References

1. D.A. Winter, *Biomechanics and Motor Control of Human Movement* (Wiley, Hoboken, 2009) ISBN 9780470398180

2. A.S. Buchman, P.A. Boyle, S.E. Leurgans, L.L. Barnes, D.A. Bennet, Cognitive function is associated with the development of mobility impairments in community-dwelling elders. Am. J. Geriatr. Psychiatr. **190**(6), 571–580 (2011). doi:10.1097/JGP.0b013e3181ef7a2e.Cognitive
3. H. Hatze, Neuromusculoskeletal control systems modeling–a critical survey of recent developments. IEEE Trans. Autom. Control, **250**(3), 375–385 (1980), ISSN 0018-9286. doi:10.1109/TAC.1980.1102380
4. J.C. Waterland, in *Integration of Movement*, Biomechanics I 1st International Seminar (S. Karger, Zurich, 1968), pp. 178–187
5. P.A. Guertin, The mammalian central pattern generator for locomotion. Brain Res. Rev. **620**(1), 45–56 (2009)
6. A.H. Cohen, in *Control Principles for Locomotion Looking Toward Biology*, Adaptive Motion of Animals and Machines, (Springer, Tokyo, 2006), pp. 41–51
7. V. Medved, in *Methodological Background*, Measurement of Human Locomotion, (CRC Press, Boca Raton, 2000) pp. 15–35
8. W.S. Harwin, J.L. Patton, V. Reggie Edgerton, Challenges and opportunities for robot-mediated neurorehabilitation. Proc. IEEE, **940**(9), 1717–1726 (2006), ISSN 00189219. doi:10.1109/JPROC.2006.880671
9. J.M. Potter, A.L. Evans, G. Duncan, Gait speed and activities of daily living function in geriatric patients. Arch. Phys. Med. Rehabil. **760**(November), 997–999 (1995), ISSN 00039993. doi:10.1016/S0003-9993(95)81036-6
10. B.B. Johansson, Current trends in stroke rehabilitation. a review with focus on brain plasticity. Acta Neurol. Scand. **1230**(19), 147–159 (2011), ISSN 00016314. doi:10.1111/j.1600-0404.2010.01417.x
11. Paralysis Injury International Campaign for Cures of Spinal Cord. General Information (2015). http://campaignforcure.org/
12. P.L. Ditunno, M. Patrick, M. Stineman, J.F. Ditunno, Who wants to walk? Preferences for recovery after SCI: a longitudinal and cross-sectional study. Spinal cord: Off. J. Int. Med. Soc. Parapleg. **46**, 500–506 (2008), ISSN 1362-4393. doi:10.1038/sj.sc.3102172
13. M. Bax, M. Goldstein, P. Rosenbaum, A. Leviton, N. Paneth, B. Dan, B. Jacobsson, D. Damiano, Proposed definition and classification of cerebral palsy. Dev. Med. Child Neurol. **470**(April), 571–576 (2005), ISSN 0012–1622. doi:10.1017/S001216220500112X
14. C. Cans, J. De-la Cruz, M.-A. Mermet, Epidemiology of cerebral palsy. Paediatr. Child Health, **180**(9), 393–398 (2015). doi:10.1016/j.paed.2008.05.015. http://www.paediatricsandchildhealthjournal.co.uk/article/S1751-7222(08)00132-7/abstract
15. J. Parkes, B. Caravale, M. Marcelli, F. Franco, A. Colver, Parenting stress and children with cerebral palsy: a European cross-sectional survey. Dev. Child Neurol. **53**, 815–821 (2011), ISSN 00121622. doi:10.1111/j.1469-8749.2011.04014.x
16. World Health Organization: What are the public health implications of global ageing? (2011). http://www.who.int/features/qa/42/en/
17. N.B. Alexander, A. Goldberg, Gait disorders: search for multiple causes. Clevel. Clin. J. Med. **720**(7) (2005), ISSN 08911150. doi:10.3949/ccjm.72.7.586
18. P.K. Canavan, L.P. Cahalin, S. Lowe, D. Fitzpatrick, M. Harris, P. Plummer-D'Amato, Managing gait disorders in older persons residing in nursing homes: a review of literature. J. Am. Med. Dir. Assoc. **100**(4), 230–237 (2009), ISSN 15258610. doi:10.1016/j.jamda.2009.02.008
19. P.G. MacRae, L.A. Asplund, J.F. Schnelle. A walking program for nursing home residents: effects of walk endurance, physical activity, mobility and quality of life. J. Am. Geriatr. Soc. **440**(1),175–180 (1996)
20. P. Dario, M.C. Carrozza, E. Guglielmelli, C. Laschi, A. Menciassi, A. Micera, F. Vecchi, Robotics as a future and emerging technology: biomimetics, cybernetics, and neuro-robotics in European projects. IEEE Robot. Autom. Mag. **120**(2), 29–45 (2005)
21. J.L. Pons, R. Ceres, L. Calderón, in *Wearable Robots And Exoskeletons*, Wearable Robots: Biomechatronic Exoskeletons (2008), pp. 1–5
22. H.L. Von Gierke, H.E. Keidel, W.D. Oestreicher. *Principles and Practice of Bionics* (Technical Press, London, 1970)

23. N. Vitiello, C.M. Oddo, T. Lenzi, S. Roccella, L. Beccai, F. Vecchi, M.C. Carrozza, P. Dario, in *Neuro-robotics Paradigm for Intelligent Assistive Technologies*, Intelligent Assistive Robots, (Springer, Heidelberg, 2015) pp. 1–40
24. F.W. Van Hook, D. Demonbreun, B.D. Weiss, Ambulatory devices for chronic gait disorders in the elderly. Am. Fam. Phys. **670**(8), 1717–1724 (2003). ISSN 0002-838X
25. R. Lam, Practice tips: choosing the correct walking aid for patients. Can. Fam. Phys. Médecin de famille canadien, **530**(12), 2115–2116 (2007). ISSN 1715-5258
26. R. Ceres, J.L. Pons, L. Calderón, A.R. Mesonero-Romanos, A.R. Jiménez, F. Sánchez, P. Abizanda, B. Saro, G. Bonivardo, in *Andador activo para la rehabilitación y el mantenimiento de la movilidad natural*, IMSERSO, (2004) pp. 3–8
27. K.D. Gross, Device use: walking AIDS, braces, and orthoses for symptomatic knee osteoarthritis. Clin. Geriatr. Med. **260**(3), 479–502 (2010), ISSN 1557-8623. doi:10.1016/j.cger.2010.03.007
28. C.L. Vaughan, B.L. Davis, J.C. O'connor, *Dynamics of Human Gait* (Kiboho Publishers, Cape Town, 1999), ISBN 0620235586. http://rehab.ym.edu.tw/document/motion/GaitBook-handout.pdf
29. M.W. Whittle, *Gait Analysis: An Introduction*, 4th edn. (Elsevier, Amsterdam, 2003) http://trid.trb.org/view.aspx?id=770947
30. M. Peshkin, D.A. Brown, J.J. Santos-Munné, A. Makhlin, E. Lewis, J.E. Colgate, J. Patton, D. Schwandt, in *KineAssist: a robotic overground gait and balance training device*, IEEE 9th International Conference on Rehabilitation Robotics, (2005) pp. 241–246, ISBN 0780390032. doi:10.1109/ICORR.2005.1501094
31. K.H. Seo, J.J. Lee, The development of two mobile gait rehabilitation systems. IEEE Trans. Neural Syst. Rehabil. Eng. **170**(2), 156–166 (2009), ISSN 15344320. doi:10.1109/TNSRE.2009.2015179
32. H. Bateni, B.E. Maki, Assistive devices for balance and mobility: benefits, demands, and adverse consequences. Arch. Phys. Med. Rehabil. **860**(1), 134–145 (2005)
33. A.F. Neto, A. Elias, C. Cifuentes, C. Rodriguez, T. Bastos, R. Carelli, in *Smart Walkers: Advanced Robotic Human Walking-aid Systems*, ed. by Y. Mohammed, S. Moreno, J. Kong, K. Amirat. Intelligent Assistive Robots, (Smart Walk, 2015), ISBN 978-3-319-12921-1
34. L.L. Haubert, D.D. Gutierrez, C.J. Newsam, J.K. Gronley, S.J. Mulroy, J. Perry, A comparison of shoulder joint forces during ambulation with crutches versus a walker in persons with incomplete spinal cord injury. Arch. Phys. Med. Rehabil. **870**(1), 63–70 (2006), ISSN 0003-9993. doi:10.1016/j.apmr.2005.07.311
35. J.R. Priebe, R. Kram, Why is walker-assisted gait metabolically expensive? Gait Posture, **340**(2), 265–269 (2011), ISSN 1879-2219. doi:10.1016/j.gaitpost.2011.05.011
36. D.L. Wright, T.L. Kemp, The dual-task methodology and assessing the attentional demands of ambulation with walking devices. Phys. Ther. **72**, 306–315 (1992). ISSN 0031-9023
37. J.F. Veneman, S. Dosen, N. Miljkovic, N. Jovicic, A. Veg, D.B. Popoviu, T. Keller, A device for active posture assistance during over ground gait training. In *1st International Conference on Applied Bionics and Biomechanics*, 2010
38. A. Veg, D.B. Popović, Walkaround: Mobile balance support for therapy of walking. IEEE Trans. Neural Syst. Rehabil. Eng. **160**(3), 264–269 (2008), ISSN 15344320. doi:10.1109/TNSRE.2008.918424
39. Made for Movement, NF-Walwer (2015). http://madeformovement.com/products/nf-walker
40. A. Frizera, R. Ceres, J.L. Pons, A. Abellanas, R. Raya, The Smart Walkers as Geriatric Assistive Device. The SIMBIOSIS Purpose. In *Proceedings of the 6th International Conference of the International Society for Gerontechnology*, (2008) pp. 1–6
41. A. Elias, A. Frizera, T.F. Bastos, Robotic walkers from a clinical point of view : feature-based classification and proposal of a model for rehabilitation programs, in *XIV Reunión de Trabajo en Procesamiento de la Información y, Control*, (2011), pp. 1–5
42. D.J. Reinkensmeyer, M.L. Boninger, Technologies and combination therapies for enhancing movement training for people with a disability. J. Neuro Eng. Rehabil. **90**(1), 17 (2012), ISSN 1743-0003. doi:10.1186/1743-0003-9-17. http://www.jneuroengrehab.com/content/9/1/17

43. Hocoma. Lokomat - Functional Robotic Gait Therapy (2015), http://www.hocoma.com/products/lokomat/
44. HealthSouth. HealthSouth's AutoAmbulator (2015), www.healthsouth.com
45. J.F. Veneman, R. Kruidhof, E.E.G. Hekman, R. Ekkelenkamp, E.H.F. Van Asseldonk, H. Van Der Kooij, Design and evaluation of the LOPES exoskeleton robot for interactive gait rehabilitation. IEEE Trans. Neural Syst. Rehabil. Eng. **150**(3), 379–386 (2007), ISSN 15344320. doi:10.1109/TNSRE.2007.903919
46. S.K. Banala, S.K. Agrawal, J.P. Scholz, Active Leg Exoskeleton (ALEX) for gait rehabilitation of motor-impaired patients. in *IEEE 10th International Conference on Rehabilitation Robotics, ICORR'07*, (2007), pp. 401–407, ISBN 1424413206. doi:10.1109/ICORR.2007.4428456
47. D. Aoyagi, W.E. Ichinose, S.J. Harkema, D.J. Reinkensmeyer, J.E. Bobrow, A robot and control algorithm that can synchronously assist in naturalistic motion during body-weight-supported gait training following neurologic injury. IEEE Trans. Neural Syst. Rehabil. Eng. **150**(3), 387–400 (2007), ISSN 15344320. doi:10.1109/TNSRE.2007.903922
48. J.L. Emken, R. Benitez, D.J. Reinkensmeyer, Human-robot cooperative movement training: learning a novel sensory motor transformation during walking with robotic assistance-as-needed. J. Neuroeng. Rehabil. **40**(8) (2007), ISSN 1743-0003. doi:10.1186/1743-0003-4-8. http://www.pubmedcentral.nih.gov/articlerender.fcgi?artid=1847825&tool=pmcentrez&rendertype=abstract
49. M.A. Guadagnoli, T.D. Lee, Challenge point: a framework for conceptualizing the effects of various practice conditions in motor learning. J. Mot. Behav. **36**, 212–224 (2004), ISSN 0022-2895. doi:10.3200/JMBR.36.2.212-224
50. N. Chockalingam, F. Chatterley, A.C. Healy, A. Greenhalgh, H.R. Branthwaite, Comparison of pelvic complex kinematics during treadmill and overground walking. Arch. Phys. Med. Rehabil. **930**(12), 2302–2308 (2012), ISSN 00039993. doi:10.1016/j.apmr.2011.10.022
51. J.R. Watt, J.R. Franz, K. Jackson, J. Dicharry, P.O. Riley, D.C. Kerrigan, A three-dimensional kinematic and kinetic comparison of overground and treadmill walking in healthy elderly subjects. Clin. Biomech. (Bristol, Avon) **250**(5), 444–449 (2010), ISSN 1879-1271. doi:10.1016/j.clinbiomech.2009.09.002. http://www.ncbi.nlm.nih.gov/pubmed/20347194
52. A. Frizera, R. Ceres, E. Rocon, J.L. Pons, A. Frizera-Neto, R. Ceres, E. Rocon, J.L. Pons, Empowering and Assisting Natural Human Mobility: the Simbiosis Walker. Int. J. Adv. Robot. Syst. **80**(3), 34–50 (2011). http://oa.upm.es/13856/2/INVE_MEM_2011_115583.pdf

Chapter 2
Human-Robot Interaction for Assisting Human Locomotion

Industrial robotics has been a developing area for more than 30 years. Initially, robotic applications were performed in confined workplaces: although robots were guided by humans from control panels, they did not have autonomous capabilities for cooperating directly with humans on the working space [1]. Nowadays, the use of robotics has extended from the industrial field to living and working places. Service robots for personal and domestic use include vacuum and floor cleaning, lawn-mowing robots, and entertainment and leisure robots such as toys, hobby devices, and robots for education and research. A successful example was introduced by iRobot with the vacuum cleaner robots in 2002, and more than 10 million home robots have been sold worldwide by this company [2].

In this context, intelligent service robotics is a research field that became very popular over the past years. It covers a wide range of scenarios, such as interactive guiding robots in museums [3], exhibitions [4] and shopping malls [5]. In the same manner, several research projects in many countries are focused on robots for assisting elderly and/or people with disabilities. Sales on this important future market of service robots is estimated to reach approximately 12,400 units in the period of 2014–2017 [6]. This market is expected to increase substantially within the next 20 years. In this scenario, the mobile robots are expected to cover a wide range of applications, such as hospital support [7] and humanoid assistance for elderly people [8].

In the future, applications for services robots may include medical, domestic, personal assistance home care, public-oriented service, cooperative material handling, power extenders, physical rehabilitation devices, physical training and entertainment. Due to the fact that service robots will often share their workspace with humans, and a direct contact between human and robots will be inevitable, robot developers should focus on both the robot and the user, more specifically on the interaction of both agents. This should take into account that, first, robots should be instructed as effectively and intuitively as a human [9] and, second, that the direct contact between human and robots must be detected as a control input to prevent human injuries and to guide the robot in its tasks. In this direction, researchers worldwide are studying

C.A. Cifuentes and A. Frizera, *Human-Robot Interaction Strategies for Walker-Assisted Locomotion*, Springer Tracts in Advanced Robotics 115,
DOI 10.1007/978-3-319-34063-0_2

the social factors related to the Human-Robot Interaction (HRI) in human environments and great attention is being focused on the Cognitive Human-Robot Interaction (cHRI) [1].

Previous approaches to guide mobile robots often involved the human as an obstacle, which had to be avoided in any case. In contrast, this chapter will adapt concepts of human-robot interaction to propose new strategies, in which a robot must behave in a assistive way, not avoiding the human, and promoting human locomotion. This could be applied in the design of new devices for functional compensation and rehabilitation of the gait according to concepts previously presented in Chap. 1.

2.1 Dual Human-Robot Interaction in Assisted Locomotion

Humans beings interact with the environment through cognitive processes, sequences of tasks that include reasoning, planing, and finally the execution of a previously identified problem or goal. From this process, the robots may use information regarding human expressions and/or physiological phenomena to adapt, learn and optimize their functions, or even to transmit back a response resulting from a cognitive process performed within the robot. This concept is named as Cognitive Human-Robot Interaction (cHRI) [10].

Considering that both agents (human and robot) share the same space, a physical Human-Robot Interaction (pHRI) may also occur. In pHRI, humans and robots share the same workspace, exchanging forces, and, possibly, cooperating.

The dual cognitive and physical interaction with humans applied in wearable robotics is explained by Pons et al. [10]. This concept can be extended in walker-assisted gait. Thus, on the one hand, the key role of a robot in a pHRI is the generation of supplementary forces to empower the human locomotion. This involves a net flux of power between both actors. On the other hand, one of the crucial roles of a cHRI is to make the human aware of the possibilities of the robot while allowing him to maintain control of the robot at all times.

In this book, control strategies based on the combination of cognitive and physical interactions in the context of human locomotion assistance are explored. Thus, physical interaction can help in setting rules for cognitive evaluations of the environment during interaction tasks. For instance, a smart walker could provide the user different levels of force feedback according to different types of therapy, or regarding inadequate gait patterns. In addition, cognitive aspects may improve the physical interaction by setting suitable control interaction parameters such as human velocity tracking (cognitive process), and reaching the desired human-robot distance and robot orientation during the walking.

In Fig. 2.1, the components involved in a cHRI are shown. In such example, a robot acts as a companion in front of the human [11]. This approach is deeply discussed in this book. This application can be useful in factories, supermarkets, libraries, restaurants or other environment where the user needs to access items to

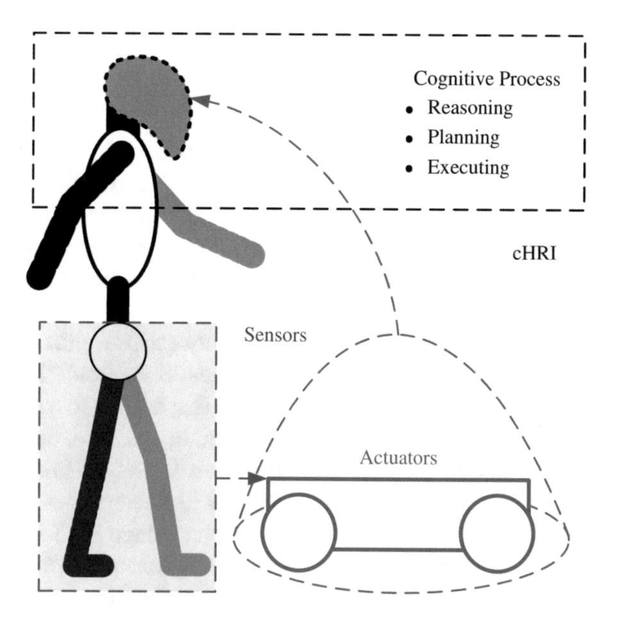

Fig. 2.1 cHRI applied in a carrier robot

be dispensed, such as tools, materials or merchandise. In such cases, the robot may act as a carrier device.

cHRI systems present often a bidirectional communication channels. On the one hand, robot's sensors measure the human actions and expressions. On the other hand, the actuators transmit the robot's cognitive information (the interpretation of the user's motion and its conversion to motor commands) to the user. In other words, the user observes the state of the system through a feedback sent immediately after the user command is executed. This configuration performs a close loop human-robot interaction in order to develop a natural cooperation during the human walking.

In Fig. 2.1, the carrier robot application does not require any physical contact with the human to be guided. However, physical contact situations should be integrated into the control law or, at least, considered in the safety requirements to avoid risks to the user. Additionally, the integration of force interaction may allow the system to assist an elderly or fragile person who needs body-weight support to walk. This application describes the scope of smart walkers as an assistive device, as aforementioned in the previous chapter. That way, this configuration could integrate concepts of HRI in the smart walkers field. This configuration will be addressed in the Chap. 4.

The combination of cHRI and pHRI could enable the development of more adaptable and safer robotic walkers, which could be beneficial to improve the human-walker interaction. This concept can be applied in different walker frames to improve the locomotion capacities as it can be seen in Fig. 2.2. In Fig. 2.2a, the carrier robot configuration provides partial body-support during the walking. In the same manner, this concept can be applied in an overground walking rehabilitation device as it can be seen in Fig. 2.2b.

2.2 General Aspects of Human-Robot Interfaces for Smart Walkers

Humans perceive the environment in which they live through their senses, such as vision, hearing, touch, smell and taste. They act on the environment using their actuators, e.g. muscles, to control body segments, hands, face, and voice. Human-to-human interaction is based on sensory perception of actuator actions. A natural communication among humans also involves multiple and concurrent modes of communication [12].

The goal of effective interaction between user and robot assistant makes it essential to provide a number of broadly utilizable and potentially redundant communication channels. This way, any HRI system that aspires to have the same naturalness should be multimodal. Different sensors can, in that case, be related to different communication modalities [12]. The integration of classic Human-Computer interfaces (HCi) like graphical input-output devices, with newer types of interfaces, such as speech or visual interfaces, tactile sensors, Laser Range Finder Sensors (LRF), Inertial Measurement Units (IMU) and force/torque sensors, facilitates this task [9].

In order to propose a multimodal interface based on the combination of cognitive and physical interaction to assist the human locomotion, this section will discuss different modalities that have been broadly used for robotics. At the same time, some relevant works related to smart walkers and sensor devices will be addressed. These sensor modalities are grouped in two categories: cHRi and pHRi.

2.2.1 Cognitive Human-Robot Interfaces

A cognitive Human-Robot interface (cHRi) is explicitly developed to support the flow of information in the cognitive interaction (possibly two-way) between the robot and

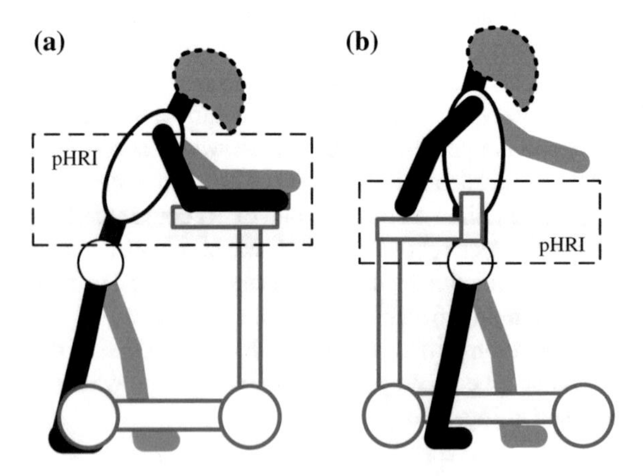

Fig. 2.2 Applications of pHRI. **a** Functional compensation of the gait. **b** Overground walking rehabilitation

the human. Information is the result of processing, manipulating and organizing data, and so the cHRi in the human-robot direction is based on data acquired by a set of sensors to measure bioelectrical and biomechanical variables [10]. Consequently, some cognitive sensor interfaces that could be useful in mobility assistance will be shown in this section: audio sensing, visual sensing, active ranging sensing and full-body motion capturing. It is important to state that, although very important in other rehabilitation and functional compensation scenarios, bioelectrical signals are not taken into account in this book, and the interfaces here presented are focused on the interaction between a human and a robotic walker.

2.2.1.1 Audio Sensing

Human voice is a natural way of communication. Although the development of a voice controlled robot system could be useful in HRI applications, this communication channel could be slow for specific human-robot scenarios, such as emergency situations or specific cases that requires a fast human reaction.

A voice recognition strategy to command the navigation of mobile robot systems over specific environment conditions was developed in [13]. However, according to the authors' knowledge, there are no reported successful cases in the field of mobility assistance. Thus, a robust speech recognition and understanding is still a research topic. Additionally, the recognition process in noisy environments has to be addressed, and also in the understanding of the meaning of the words. The robot and the human must have a common understanding of the situation [9], which can be very complex considering the wide range of scenarios for real-life mobility devices.

2.2.1.2 Visual Sensing

From a practical standpoint, visual sensing involves the processing of a great amount of information in real time, which could put undue demands on the processing power of the system being controlled. Furthermore, visual sensing requires an unoccluded view of the human, putting restrictions on the motion of the user and the physical setting for HCI [12].

Nonetheless, a video camera, along with a set of techniques for processing and interpreting the image sequence, may allow the incorporation of a variety of human-actions modalities into HRI through visual sensing. These actions may include hand gestures for applications, such as the ability of distinguishing the gesture from no sign [14], a human pointing at objects or locations of interest to the robot, or an autonomous robot asking for directions from humans and interpreting those directions [15].

Gaze direction normally indicates a person's interest in his/her surrounding, so it can be exploited as a very easy way to tell the robot what a user wants [16]. Additionally, head gestures recognition has been used to guide movements of an intelligent wheelchair [17], or body movements capturing as an interface to be used in non-specific environments [18]. As it can be observed, gesture and mimic recognition is an ongoing research activity in the fields of human-computer and the human-robot dialog.

Recently, there are some related works that use visual sensing for smart walkers in indoor environments. For instance, in [19], for monitoring and control purposes, a fast feet position and orientation detection algorithm is proposed. It is based on an on-board camera with a depth sensor and does not require the use of any marker. Moreover, a robotic walker that localizes the user, estimates the body pose and recognizes human actions, gestures and intentions was presented in [20].

2.2.1.3 Active Ranging Sensing

A Laser Range Finder (LRF) is a time-of-flight sensor that achieves significant improvements over the ultrasonic range sensor as perform good measurement precision and accuracy in a planar range of measurement. This type of sensor consists of a emitter that illuminates a target with a laser beam, and a receiver capable of detecting the component of light returned. These devices produce a range estimation based on the time needed for the light to reach the target and return to the sensor device. In most commercial devices, the light beam rotates in mirror and sweeps through a mechanic device to cover a target plane [21].

Human tracking is essential for mobile service robots and human-robot interaction applications. There are a variety of approaches, and most of them employ both visual sensing and/or LRF devices [22–25]. However, when the robot is tracking a person in outdoors, visual measurement errors are expected to increase. For this reason, some researches apply LRF human tracking [26–30]. The use of LRF is advantageous because it is robust with respect to illumination changes in the environment.

One common way for human detection by LRF is scanning the individual's legs. In this case, apart from tracking the position of the human in relation to the robot, other important human gait information can be obtained allowing a more adaptable human-robot interaction. Step length, cadence, velocities, legs orientation, and gait phases (stance and swing) are some examples of information that may be obtained from tracking the human legs. Nevertheless, it is important to observe that the tracking system has to deal with specific situations, such as clothing. Therefore the use of some clothes that fully covers user's legs, such as long skirts, is a limitation and it is not considered in this book.

A basic technique for leg detection uses the acquired data from LRF, defining the measurement range that violates the static environment assumption to determine the leg position [25]. Other approaches make use of specific geometrical shapes to determine the leg position. In [26], circle shapes are suggested to extract leg data. In [27, 28], the approach is inductive on the basis of sufficient measurements without specific assumptions of shapes and also exploiting a human walking model. Such approaches do not present an exhaustive experimental evaluation and do not explain how the performance of the detection algorithm is affected when the legs cannot be detected. It is important to understand that leg obstruction is very common in curved paths, when one leg may be placed behind the other from the sensor point-of-view.

In the field of Smart Walkers, legs tracking approach has been also implemented; some relevant examples are [31–33]. In [31], two infrared sensors were used and, in [32, 33], two LRF sensors were used to perform the scanning of each leg. This interface does not require the user to produce any specific trained command to generate walker motion.

Another proposal suggests the use of LRF for human torso tracking [29, 30]. An advantage of this approach is that the scanned data presents smaller variations caused by the oscillatory movements during the human gait, and obstruction and occlusion issues when performing curved paths do not represent a problem. As a disadvantage, human gait information (a fully modeled process) is not measured and cannot be used as an extra input to the system. The works presented in [29, 30] propose the estimation of body pose information using particle filters. However, the human tracking is not effective when detecting non-human objects with similar shape and width of human segments.

2.2.1.4 Full-Body Motion Capturing

Several systems are available to measure the motion of the human body. Traditionally, such systems are based on three-dimensional photogrammetry, which are considered a gold standard for human movement analysis due to the great precision. However, these systems present some important disadvantages related to the high cost, occlusion of markers by the body or external elements [34] and, most importantly, low portability to be used in HRI applications.

The popularization of MicroElectroMechanical Systems (MEMS) had an important impact in several sectors of the industry and in the research community [35, 36]. Such elements are integrated into many devices for final consumers, such as laptops, mobile phones, entertainments, and mobile robots [37]. Developments of small and light inertial sensors based on MEMS have allowed their use on human body limbs without interfering in natural movements of the user [38, 39].

Body segment orientation can be estimated by using the combination of different sensors through data fusion techniques. Usually, accelerometers (inclination), gyroscopes (angular velocity), magnetometers (heading angle), and temperature sensors (for thermal drift compensation) are used together by means of fusion algorithms [40]. This composes an inertial sensor or IMU (Inertial Measurement Unit). The combination of linear accelerations, angular velocities and the reference of the Earth's magnetic field allows the measurement of tridimensional orientation of the device. By placing IMUs in different body segments, it is possible to obtain a complete description of the human joint kinematics during the execution of different tasks. Finally, considering the recent evolution of the communication devices, it is possible to build IMUs that can also transmit those measurements wirelessly [41, 42].

In addition, the development of wearable IMU systems presents important advantages in the field of human motion capturing: portability, high accuracy and ease of use in unstructured environments. The integration of the wearable sensors and mobile robots is expected to enable a new generation of service robots and healthcare applications [43]. Wearable IMU sensors are appearing in this field, offering the possibility of combining human tracking with human gesture detection and body posture estimation [44–46]. The next section addresses some approaches that combine LRF and IMU sensors in mobile robotics applications.

2.2.1.5 Human Tracking: LRF and IMU Sensors

As previously mentioned, the human tracking performed by a LRF sensor is often not effective when detecting non-human objects with similar shape and width of human segments. In addition, the use of wearable IMU sensors (already fully integrated in personal mobile devices) on the human's body may present important advantages by eliminating the possibility of uncertain situations regarding LRF sensors. Consequently, the combination of LRF and wearable IMU sensors are appearing in this field as it offers the possibility of combining human tracking with human gesture detection and body posture estimation [44–46].

Considering the combination of LRF and IMU sensors for human tracking, in [44], a method for combining kinematic measurements from a LRF mounted on the robot and an IMU carried by the human is shown. A proposal to extract human velocity and position is also presented. However, that study does not provide any information regarding the validation of the proposed method. In [45], a study in which several robots were programmed to follow a person for the purpose of mapping a building for firefighters' rescue missions is presented. This sensor combination is employed to avoid the use of information obtained from artificial vision systems, such as cameras. In this case, the objective is to map the building for situations in which there is low visibility caused by fire. An IMU was used for mapping and locating the current position of a firefighter and, finally, providing the subject an exit path. Finally, a method for human motion capturing in large areas is described in [46], which shows a tracking approach that aims to provide globally aligned full body posture estimates by combining information from sensor on a mobile robot and multiple wearable IMU sensors attached to the human subject.

Summarizing, works found in the literature indicate a trend for future developments in the field of human tracking using mobile robots that rely on the integration of LRF and human motion capturing by means of wearable IMU sensors. This approach needs further investigation, and appropriate sensor integration algorithms have to be implemented, which is the one of the focuses of the work presented in this book.

2.2.2 Physical Human Robot Interfaces

A physical Human-Robot interface (pHRi) is based on a set of actuators and a rigid structure that is used to transmit forces to the human musculoskeletal system. The close physical interaction through this interface imposes strict requirements on robots as regards safety and dependability [10]. Some sensors that may be used in physical interfaces related to mobility assistance will be presented in this section: position and motion sensing, tactile and force sensing and also some issues regarding force signal processing to control the motion of robotic walkers.

2.2.2.1 Position and Motion Sensing

A large number of interface devices have been built to sense the position and motion of the human hand and other body parts for use in HCI. One of the simplest variations of such interface is the keyboard, where the touch of a particular key indicates that one of a set of possible inputs was selected. More accurate position and motion sensing in a 2-D plane is used in interface devices such as a mouse, light pen, stylus and tactile displays. Three dimensional position/motion sensing is commonly done through a joystick, a trackball or hand glove devices [12]. In human mobility assistance, conventional interfaces such as buttons [47, 48], joysticks [49] and touch screens [50] have been used to directly guide a robotic walker.

These interfaces present important delays and loss of information in the conversion of human movements into discrete and unnatural events. In that case, some issues can be analyzed, such as the information that may be lost in the translation of the human-intentions task into discrete events, delays that are introduced when natural cognitive process are encoded into and imposed to sequential task and the necessary training phase that is needed to teach the user to generate no natural commands [51].

2.2.2.2 Tactile and Force Sensing

Direct physical contact represents undoubtedly the most subtle and critical form of interaction between humans and machines. Any motion of a machine, which occurs in close proximity to a human, and any force exerted by the robot has to be soft and compliant and must never exceed the force exerted by the human to protect her/him [9].

Handlebars, as an alternative, are a common way of providing not only guidance but also weight support in mobility assistance devices. Some approaches [52–54] integrate pressure and force sensors into handlebars to get user's movement intention. Other approaches replace the handlebars with forearm supporting platforms, which allow a better posture and stability during gait. Usually, a sensor interface is integrated inside this platform, such as a joystick [55], a 3D force sensor [56] or two 3D force sensors (one for each forearm support) [57]. This integration presents a more natural way to command the walker motion without previous training. Some remarks regarding the processing of interaction force signals during walker-assisted gait will be addressed in the next section.

2.2.2.3 Extraction of Upper-Limbs Guiding Intentions

In previous works [57–59], the components of upper limb reaction forces during the walker-assisted gait were identified and characterized. These approaches presented preliminary human-walker trials performed without traction or controlled motions, which were useful to detect the presence of noise or elucidate undesired components that could affect the control strategy. This integration presents a more natural way to command the walker motion without previous training.

In [58], a study showed that vertical components of forces are highly correlated with gait phases. This work identified that the gait cadence and the user's partial body weight are represented for independent components contained into each force signal. A methodology to extract user's navigation commands related to components from upper-body force interaction data was presented in [57, 59]. In these studies, a low-pass filter is used to eliminate the frequency components introduced by ground-wheel interaction and an adaptive notch filter was implemented to reject the interaction force components caused by the user's trunk motion during the gait. These components are present due to the natural trunk oscillations caused by the alternated supports and do not reflect the desired navigation commands. These filters are adjusted with an online estimation of the gait cadence. Two alternatives were evaluated as follows.

In [59], a combination of the vertical force components of each arm is used for continuous estimation of the gait cadence. This architecture was evaluated with healthy subjects, and provided a high rate of cancelation of trunk components. Even though subjects with gait disorders usually present asymmetric gait patterns, it is not possible to obtain robust cadence estimation only from a combination of the vertical force components. Thus, cadence estimation directly from the user's legs position was presented in [57], using ultrasonic sensors. A high rate of cancelation of trunk components for patients with disabilities was obtained. The main disadvantage of this approach is that the user had to wear sensors on each leg compromising the usability of the device.

2.3 Proposal of a HRI Multimodal Interface

Taking into account the concept of dual cognitive and physical HRI, a new multimodal interface for walker-assisted gait is proposed in Fig. 2.3. This interface involves the integration of different modalities to promote a natural HRI during the walking.

The multimodal interface combines active ranging sensing (LRF) and human motion capturing (IMU) to develop legs and trunk tracking. In addition, force sensing is included to obtain information regarding the interaction forces between the human and the walker. The detailed design and validation of this sensor interface and the interaction strategies in which they are integrated will be addressed in the next chapters.

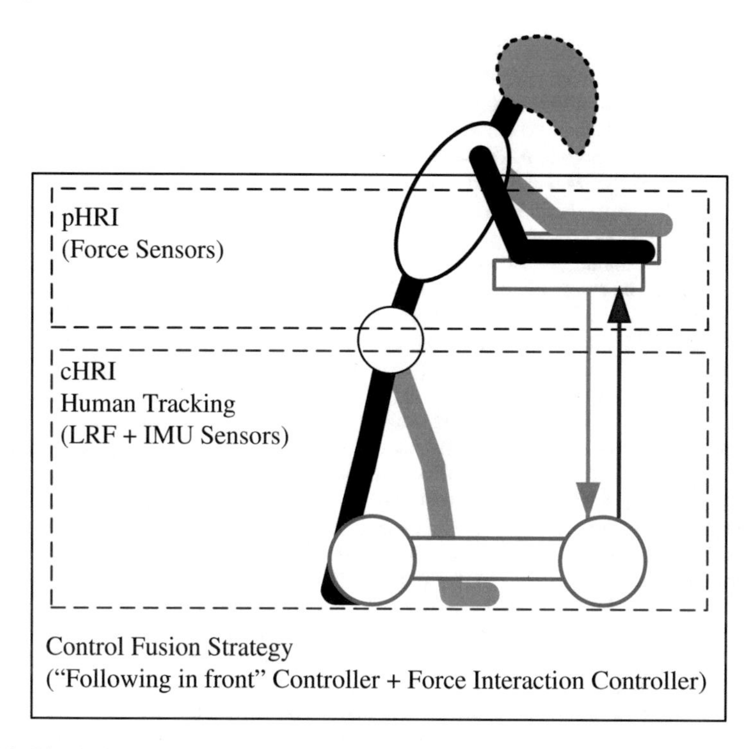

Fig. 2.3 Physical and cognitive HRI for walker-assisted gait

Two close control loops are proposed to naturally adapt the walker position and to perform body weight support strategies. On the one hand, a force interaction controller generates velocity output to the walker based on the upper-limbs interaction forces (grey arrow in Fig. 2.3). On the other hand, a controller keeps the walker within a desired position ("following in front") to the human to improve the physical interaction.

The user receives two cognitive information during the walking, such as: visual information regarding the robot following in front of the user, and the force feedback (black arrow in Fig. 2.3) related to the pHRI during partial body weight support. The formulation and implementation regarding these control strategies will be presented in the next chapters.

References

1. A. De Santis, *Modelling and control for human robot interaction*. Ph.D. thesis, Universita' Degli Studi Di Napoli Federico II Dottorato (2007). http://www.fedoa.unina.it/2067/1/De_Santis_Ingegneria_Informatica_Automatica.pdf
2. IRobot. About iRobot (2013). http://www.irobot.com/us/Company/About.aspx

3. W. Burgard, A.B. Cremers, D. Fox, D. Hähnel, G. Lakemeyer, D. Schulz, W. Steiner, S. Thrun, Experiences with an interactive museum tour-guide robot. Artif. Intell. **1140**(1–2), 3–55 (1999). ISSN 00043702 . doi:10.1016/S0004-3702(99)00070-3. http://linkinghub.elsevier.com/retrieve/pii/S0004370299000703
4. P. Trahanias, W. Burgard, A. Argyros, D. Hahnel, H. Baltzakis, P. Pfaff, C. Stachniss, Tourbot and WebFAIR: Web-operated mobile robots for tele-presence in populated exhibitions. IEEE Robot. Autom. Mag. **120**(2), 77–89 (2005). http://ieeexplore.ieee.org/xpls/abs_all.jsp?arnumber=1458329
5. T. Kanda, M. Shiomi, Z. Miyashita, H. Ishiguro, N. Hagita. An affective guide robot in a shopping mall, in: *Proceedings of the 4th ACM/IEEE international conference on Human robot interaction - HRI '09*, p. 173, ACM Press, New York, USA (2009). ISBN 9781605584041. doi:10.1145/1514095.1514127. http://portal.acm.org/citation.cfm?doid=1514095.1514127
6. IFR International Federation of Robotics. Service Robot Statistics (2012). http://www.ifr.org/service-robots/statistics/
7. J. Hu, A. Edsinger, N. Donaldson, M. Solano, A. Solochek, R. Marchessault. An advanced medical robotic system augmenting healthcare capabilities - robotic nursing assistant, in: *Proceedings of the 2011 IEEE International Conference on Robotics and Automation* (IEEE, 2011), pp. 6264–6269. ISBN 978-1-61284-386-5. doi:10.1109/ICRA.2011.5980213. http://ieeexplore.ieee.org/lpdocs/epic03/wrapper.htm?arnumber=5980213
8. Z. Mohamed, G. Capi, Development of a New Mobile Humanoid Robot for Assisting Elderly People. *Procedia Engineering*, 410 (Iris): 345–351 (2012). ISSN 18777058. doi:10.1016/j.proeng.2012.07.183. http://linkinghub.elsevier.com/retrieve/pii/S1877705812025696
9. B. Siciliano, O. Khatib, F. Groen, *Springer Tracts in Advanced Robotics Volume 14 Springer Tracts in Advanced Robotics*, vol. 14 (Springer, Berlin, 2005). ISBN 3540232117
10. J. Pons, R. Ceres, L. Calderón, Chapter Introduction to Wearable Robots and Exoskeletons, in *Wearable Robots: Biomechatronic Exoskeletons*, Wiley, pp. 1–5 (2008)
11. D. M Ho, J.-S. Hu, J.-J. Wang, Behavior control of the mobile robot for accompanying in front of a human, in *Proceedings of the 2012 IEEE/ASME International Conference on Advanced Intelligent Mechatronics (AIM)*, pp. 377–382 (2012). doi:10.1109/AIM.2012.6265891. http://ieeexplore.ieee.org/lpdocs/epic03/wrapper.htm?arnumber=6265891
12. R. Sharma, V.I. Pavlovic, T.S. Huang, Toward multimodal human-computer interface, in *Proceedings of the IEEE*, 860(5): 853–869 (1998). ISSN 00189219. doi:10.1109/5.664275. http://ieeexplore.ieee.org/lpdocs/epic03/wrapper.htm?arnumber=664275
13. H.C. Nguyen, S.B. Kyun, C. Kang. Integration of robust voice recognition and navigation system on mobile robot, in *2009 ICROS-SICE International Joint Conference*, pp. 2103–2108 (2009) http://ieeexplore.ieee.org/xpls/abs_all.jsp?arnumber=5333843
14. R.C. Luo, Y.-C. Wu, Hand gesture recognition for Human-Robot Interaction for service robot, in *2012 IEEE International Conference on Multisensor Fusion and Integration for Intelligent Systems (MFI)*, pp. 318–323. (IEEE, 2012). ISBN 978-1-4673-2512-7. doi:10.1109/MFI.2012.6343059. http://ieeexplore.ieee.org/lpdocs/epic03/wrapper.htm?arnumber=6343059
15. M. Van den Bergh, D. Carton, R. De Nijs, N. Mitsou, C. Landsiedel, K. Kuehnlenz, D. Wollherr, L. Van Gool, M. Buss, Real-time 3D hand gesture interaction with a robot for understanding directions from humans, in *2011 Ro-Man*, pp. 357–362. (IEEE, 2011). ISBN 978-1-4577-1571-6. doi:10.1109/ROMAN.2011.6005195. http://ieeexplore.ieee.org/lpdocs/epic03/wrapper.htm?arnumber=6005195
16. R. Atienza, A. Zelinsky, Active gaze tracking for human-robot interaction, in *Proceedings. Fourth IEEE International Conference on Multimodal Interfaces*, IEEE Comput. Soc., pp. 261–266 (2002). ISBN 0-7695-1834-6. doi:10.1109/ICMI.2002.1167004. http://ieeexplore.ieee.org/lpdocs/epic03/wrapper.htm?arnumber=1167004
17. Zhang-fang Hu Lin Li, Y. Luo, Y. Zhang, X. Wei. A novel intelligent wheelchair control approach based on head gesture recognition, in *2010 International Conference on Computer Application and System Modeling*, number Iccasm, pp. 159–163 (2010). ISBN 9781424472376
18. J. Fernandez, J. Aranda, Automatic Control, Computer Engineering Dpt, and Universitat Politecnica De Catalunya. Visual Human Machine Interface by Gestures, in *Proceedings of the*

2003 IEEE International Conference on Robotics and Automation, pp. 386–391 (2003). ISBN 0780377362
19. S. Page, M.M. Martins, L. Saint-bauzel, C.P. Santos, V. Pasqui. Fast embedded feet pose estimation based on a depth camera for smart walker, in *Proceedings of the IEEE Conference on Robotics and Automation - ICRA* (2015)
20. G. Chalvatzaki, G. Pavlakos, K. Maninis, X. Papageorgiou, V. Pitsikalis, C. Tzafestas, P. Maragos, Towards an intelligent robotic walker for assisted living using multimodal sensorial data, in *4th International Conference on Wireless Mobile Communication and Healthcare - Transforming Healthcare Through Innovations in Mobile and Wireless Technologies*, pp. 156–159 (2014). ISBN 978-1-63190-014-3. doi:10.4108/icst.mobihealth.2014.257358. http://eudl.eu/doi/10.4108/icst.mobihealth.2014.257358
21. R. Siegwart, I.R. Nourbakhsh, in *Introduction to Autonomous Mobile Robots* (2004). ISBN 026219502X. http://books.google.com/bookshl=en&lr=&id=gUbQ9_weg88C&oi=fnd&pg=PR11&dq=Introduction+to+Autonomous+Mobile+Robots&ots=X3G6WEu86O&sig=AjlpAnd9od0OFbqsfnXEv3f89YY
22. N. Bellotto, H. Hu, Multisensor-based human detection and tracking for mobile service robots. IEEE Trans. Syst. Man Cybern. **390**(1), 167–181 (2009). http://ieeexplore.ieee.org/xpls/abs_all.jsp?arnumber=4695975
23. V Alvarez-santos, X M Pardo, R Iglesias, A Canedo-rodriguez, and C V Regueiro. Feature analysis for human recognition and discrimination : Application to a person-following behaviour in a mobile robot. *Robotics and Autonomous Systems*, 600 (8):0 1021–1036, 2012. ISSN 0921-8890. doi:10.1016/j.robot.2012.05.014
24. Y. Motai, S.K. Jha, D. Kruse, Human tracking from a mobile agent: optical flow and Kalman filter arbitration. Signal Process. Image Commun. **270**(1), 83–95 (2012). ISSN 09235965 doi:10.1016/j.image.2011.06.005. http://linkinghub.elsevier.com/retrieve/pii/S0923596511000713
25. D.-H. Kim, Y. Lee, J.-Y. Lee, G.-J. Park, C.-S. Han, S.K. Agrawal, Detection, motion planning and control of human tracking mobile robots, in *Proceedings of the 2011 8th International Conference on Ubiquitous Robots and Ambient Intelligence (URAI)*, pp. 113–118. (IEEE, 2011). ISBN 978-1-4577-0723-0. doi:10.1109/URAI.2011.6145943. http://ieeexplore.ieee.org/lpdocs/epic03/wrapper.htm?arnumber=6145943
26. J. Xavier, M. Pacheco, D. Castro, A. Ruano, U. Nunes, Fast line, arc/circle and leg detection from laser scan data in a player driver. In *Proceedings of the 2005 IEEE International Conference on Robotics and Automation*, pp. 3930–3935 (2005). ISBN 078038914X. http://ieeexplore.ieee.org/xpls/abs_all.jsp?arnumber=1570721
27. Y. Sung, W. Chung, Human tracking of a mobile robot with an onboard LRF (Laser Range Finder) using human walking motion analysis, in *2011 8th International Conference on Ubiquitous Robots and Ambient Intelligence (URAI)*, vol. 1, pp. 366–370. (IEEE, 2011). ISBN 978-1-4577-0723-0. doi:10.1109/URAI.2011.6145998. http://ieeexplore.ieee.org/lpdocs/epic03/wrapper.htm?arnumber=6145998
28. W. Chung, H. Kim, Y. Yoo, C.-B. Moon, J. Park, The detection and following of human legs through inductive approaches for a mobile robot with a single laser range finder. IEEE Trans. Ind. Electron. **590**(8), 3156–3166 (2012). ISSN 0278-0046. doi:10.1109/TIE.2011.2170389. http://ieeexplore.ieee.org/lpdocs/epic03/wrapper.htm?arnumber=6032092
29. D.F. Glas, T. Miyashita, H. Ishiguro, N. Hagita, Laser tracking of human body motion using adaptive shape modeling, in *Proceedings of the 2007 IEEE/RSJ International Conference on Intelligent Robots and Systems*, pp. 602–608. (IEEE, 2007). ISBN 978-1-4244-0911-2. doi:10.1109/IROS.2007.4399383. http://ieeexplore.ieee.org/lpdocs/epic03/wrapper.htm?arnumber=4399383
30. E.-J. Jung, B.-J. Yi, S. Yuta, Control algorithms for a mobile robot tracking a human in front, in *Proceedings of the 2012 IEEE/RSJ International Conference on Intelligent Robots and Systems*, pp. 2411–2416. (IEEE, 2012). ISBN 978-1-4673-1736-8. doi:10.1109/IROS.2012.6386200. http://ieeexplore.ieee.org/lpdocs/epic03/wrapper.htm?arnumber=6386200
31. G. Lee, T. Ohnuma, N.Y. Chong, Design and control of JAIST active robotic walker. Intell. Ser. Robot. **30**(3), 125–135 (2010). ISSN 1861-2776. doi:10.1007/s11370-010-0064-5. http://link.springer.com/10.1007/s11370-010-0064-5

32. G. Lee, T. Ohnuma, N.Y. Chong, S.G. Lee, Walking intent-based movement control for JAIST active robotic walker. IEEE Trans. Syst. Man Cybern. Syst. **440**(5), 665–672 (2014). ISSN 10834427. doi:10.1109/TSMC.2013.2270225
33. T. Ohnuma, G. Lee, N.Y. Chong, Particle filter based lower limb prediction and motion control for JAIST active robotic walker, in *IEEE International Symposium on Robot and Human Interactive Communication* (2014). ISBN 9781479967650
34. J. Han, H.S. Jeon, B.S. Jeon, K.S. Park, Gait detection from three dimensional acceleration signals of ankles for the patients with Parkinson's disease, in *5th International IEEE EMBS Special Topic Conference on Information Technology in Biomedicine*, pp. 1–4 (2006). http://medlab.cs.uoi.gr/itab2006/proceedings/biosignalanalysis/108.pdf
35. R. Raya, J.O. Roa, E. Rocon, R. Ceres, J.L. Pons, Wearable inertial mouse for children with physical and cognitive impairments. Sens. Actuators Phys. **1620**(2), 248–259 (2010). ISSN 09244247. doi:10.1016/j.sna.2010.04.019. http://linkinghub.elsevier.com/retrieve/pii/S0924424710001950
36. J. Music, M. Cecic, M. Bonkovic, Testing inertial sensor performance as hands-free human-computer interface. WSEAS Trans. Comput. **80**(4), 715–724 (2009). http://www.wseas.us/e-library/transactions/computers/2009/29-182.pdf
37. B. Barshan, H.F. Durrant-Whyte, Inertial navigation systems for mobile robots. IEEE Trans. Robot. Autom. **110**(3), 328–342 (1995). ISSN 1042296X. doi:10.1109/70.388775. http://ieeexplore.ieee.org/lpdocs/epic03/wrapper.htm?arnumber=388775
38. H.J. Luinge, P.H. Veltink, Measuring orientation of human body segments using miniature gyroscopes and accelerometers. Med. Biol. Eng. Comput. **430**(2), 273–82 (2005). ISSN 0140-0118. http://www.ncbi.nlm.nih.gov/pubmed/15865139
39. G. a Hansson, P. Asterland, N.G. Holmer, S. Skerfving, Validity and reliability of triaxial accelerometers for inclinometry in posture analysis. Med. Biol. Eng. Comput. **390**(4) 405–13 (2001). ISSN 0140-0118. http://www.ncbi.nlm.nih.gov/pubmed/11523728
40. H.M. Schepers, D. Roetenberg, P.H. Veltink, Ambulatory human motion tracking by fusion of inertial and magnetic sensing with adaptive actuation. Med. Biol. Eng. Comput. **480**(1), 27–37 (2010). ISSN 1741-0444. doi:10.1007/s11517-009-0562-9. http://www.pubmedcentral.nih.gov/articlerender.fcgi?artid=2797438&tool=pmcentrez&rendertype=abstract
41. V. van Acht, E. Bongers, N. Lambert, R. Verberne, Miniature wireless inertial sensor for measuring human motions, in *Conference Proceedings: ... Annual International Conference of the IEEE Engineering in Medicine and Biology Society. IEEE Engineering in Medicine and Biology Society. Conference*, vol. 2007, pp. 6279–6282 (2007). ISBN 1424407885. doi:10.1109/IEMBS.2007.4353790. http://www.ncbi.nlm.nih.gov/pubmed/18003456
42. M. El-Gohary, L. Holmstrom, J. Huisinga, E. King, J. McNames, F. Horak. Upper limb joint angle tracking with inertial sensors, in *Conference Proceedings: ... Annual International Conference of the IEEE Engineering in Medicine and Biology Society. IEEE Engineering in Medicine and Biology Society. Conference*, **2011**, 5629–32 (2011). ISSN 1557-170X. doi:10.1109/IEMBS.2011.6091362. http://www.ncbi.nlm.nih.gov/pubmed/22255616
43. P. Bonato, Wearable Sensors and Systems. IEEE Eng. Med. Biol. Mag. **290**(3), 25–36 (2010). http://ieeexplore.ieee.org/xpls/abs_all.jsp?arnumber=5463017
44. L. Wu, Z. An, Y. Xu, L. Cui, Human Tracking Based on LRF and Wearable IMU Data Fusion, in *12th International Conference on Information Processing in Sensor Networks*, pp. 349–350 (2013). ISBN 9781450319591
45. L. Nomdedeu, J. Sales, E. Cervera, J. Alemany, R. Sebastia, J. Penders, V. Gazi, An Experiment on Squad Navigation of Human and Robots, in *10th International Conference on Control Automation*, pp. 17–20 (2008)
46. J. Ziegler, H. Kretzschmar, C. Stachniss, G. Grisetti, W. Burgard, Accurate human motion capture in large areas by combining IMU- and laser-based people tracking, in *2011 IEEE/RSJ International Conference on Intelligent Robots and Systems*, pp. 86–91 (IEEE, 2011). ISBN 978-1-61284-456-5. doi:10.1109/IROS.2011.6094430. http://ieeexplore.ieee.org/lpdocs/epic03/wrapper.htm?arnumber=6094430

47. G. Lacey, S. Macnamara, User involvement in the design and evaluation of a smart mobility aid. J. Rehabil. Res. Dev. **370**(6), 709–723 (2000). http://www.rehab.research.va.gov/jour/00/37/6/lacey376.htm
48. A.J. Rentschler, R. a Cooper, B. Blasch, M.L. Boninger, Intelligent walkers for the elderly: performance and safety testing of VA-PAMAID robotic walker. J. Rehabil. Res. Dev. **400**(5), 423–432 (2003). ISSN 0748-7711. http://www.ncbi.nlm.nih.gov/pubmed/15080227
49. H. Hashimoto, A. Sasaki, Y. Ohyama, C. Ishii, Walker with hand haptic interface for spatial recognition, in *Proceedings of the 9th IEEE International Workshop on Advanced Motion Control, 2006.*, pp. 311–316 (IEEE, 2006). ISBN 0-7803-9511-1. doi:10.1109/AMC.2006.1631677. http://ieeexplore.ieee.org/lpdocs/epic03/wrapper.htm?arnumber=1631677
50. B. Graf, M. Hans, R.D. Schraft, Care-O-bot II development of a next generation robotic home assistant. Auton. Robots **160**(2), 193–205 (2004). ISSN 0929-5593. doi:10.1023/B:AURO.0000016865.35796.e9. http://link.springer.com/10.1023/B:AURO.0000016865.35796.e9
51. L. Bueno, F. Brunetti, A. Frizera, J.L. Pons, Human-robot cognitive interaction, in *Wearable Robots: Biomechatronic Exoskeletons*, vol. 1, Chap. 4, pp. 87–126. (Wiley, 2008). ISBN 9780470512944. http://books.google.com/books?hl=en&lr=&id=ovCkTEKEmkkC&oi=fnd&pg=PA87&dq=Human+?+robot+cognitive+interaction&ots=NxheH_YNqB&sig=37faBYJkKuJTLXri_4ILqei4c5U
52. G. Lacey, D. Rodriguez-Losada, The evolution of guido. IEEE Robot. Autom. Mag. **150**(4), 75–83 (2008). http://ieeexplore.ieee.org/xpls/abs_all.jsp?arnumber=4658323
53. K.-T. Yu, C.-P. Lam, M.-F. Chang, W.-H. Mou, S.-H. Tseng, L.-C. Fu, An interactive robotic walker for assisting elderly mobility in senior care unit, in *Proceedings of the IEEE Workshop on Advanced Robotics and its Social Impacts*, pp. 24–29 (IEEE, 2010). ISBN 978-1-4244-9122-3. doi:10.1109/ARSO.2010.5679631. http://ieeexplore.ieee.org/lpdocs/epic03/wrapper.htm?arnumber=5679631
54. M.-F. Chang, W.-H. Mou, C.-ke Liao, L.-C. Fu, Design and implementation of an active robotic walker for Parkinson's patients, in *Proceedings of the SICE Annual Conference*, pp. 2068–2073 (2012)
55. M. Martins, C. Santos, A. Frizera, R. Ceres, Real time control of the ASBGo walker through a physical human-robot interface. Meas. J. Int. Meas. Conf. **480**(1), 77–86 (2014). ISSN 02632241. doi:10.1016/j.measurement.2013.10.031
56. O. Jr Chuy, Y. Hirata, Z. Wang, K. Kosuge, Motion control algorithms for a new intelligent robotic walker in, in *Proceedings of the IEEE International Conference on Mechatronics and Automation*, pp. 1509–1514 (2005). ISBN 078039044X
57. A. Frizera, R. Ceres, E. Rocon, J.L. Pons, A. Frizera-Neto, R. Ceres, E. Rocon, J.L. Pons, Empowering and assisting natural human mobility: the simbiosis walker. Int. J. Adv. Robot. Syst. **80**(3), 34–50 (2011). http://oa.upm.es/13856/2/INVE_MEM_2011_115583.pdf
58. A. Abellanas, A. Frizera, R. Ceres, J.A. Gallego, Estimation of gait parameters by measuring upper limb-walker interaction forces. Sens. Actuators Phys. **1620**(2), 276–283 (2010). ISSN 0924-4247
59. A.F. Neto, J.A. Gallego, E. Rocon, Extraction of user's navigation commands from upper body force interaction in walker assisted gait. BioMed. Eng. OnLine, **90**(37) (2010). ISSN 1475-925X. doi:10.1186/1475-925X-9-37. http://www.pubmedcentral.nih.gov/articlerender.fcgi?artid=2924341&tool=pmcentrez&rendertype=abstract, http://www.biomedcentral.com/content/pdf/1475-925X-9-37.pdf

Chapter 3
Development of a Cognitive HRI Strategy for Mobile Robot Control

The concept of a physical and cognitive HRI for walker-assisted gait was presented in the previous chapter (see Fig. 2.3). The HRI is implemented by means of a multimodal interface, which is used to develop a natural human-robot interaction in the context of human mobility assistance. That way, both cHRI and pHRI were included in this interface according to Fig. 2.3. Specifically, this chapter describes the cHRI component, which combines two sensor modalities: active ranging sensing (LRF) and human motion capturing (IMU) to perform the human tracking. This sensor combination presents important advantages to monitor the human gait from a mobile robot point of view, such as mentioned in the previous last chapter.

Moreover, the cHRI links the human tracking information with the control strategy, which could enable the robot to follow in front of the user without any contact during locomotion (as proposed in the previous chapter). That way, the cHRI block functionality can be represented as the carrier robot configuration (see Fig. 2.1). Consequently, this strategy will be evaluated in a mobile robot in order to achieve a natural "following in front of" the user. This study is based on a previous work [1]. The next chapter will address some issues regarding physical contact during the walker-assisted locomotion.

Control strategies for mobile robots following behind a user is a common approach in many works found on the literature [2, 3]. Alternatively, there are other approaches with the "side by side" behavior [4, 5]. Recently, an alternative behavior was introduced in [6, 7], where the mobile robot follows the user while positioned in front of him/her. This approach was previously presented in Fig. 2.1 as a example of a cognitive HRI model. As previously mentioned, accompanying in front of a human is useful in many applications: if the robot carries tools, materials or merchandise to be dispensed, it is more natural and comfortable for the person to access the items if the robot is placed in front of him/her [7].

Specifically in [6], the authors developed one experiment with subjects walking or running along a straight line, and a mobile robot tracking and following the subject from behind. This experiment has indicated that a robot moving behind the human

C.A. Cifuentes and A. Frizera, *Human-Robot Interaction Strategies for Walker-Assisted Locomotion*, Springer Tracts in Advanced Robotics 115,
DOI 10.1007/978-3-319-34063-0_3

causes the human to always pay attention to its motion. Therefore, the user is more comfortable when the robot accompanies staying in his/her field-of-view.

There are fundamental differences in motion between conventional wheeled mobile robots and humans. A possible solution is to use the control system to absorb this kinematic difference between human and the mobile robot locomotion. In [8, 9], a virtual spring model is used. This method is derived from the assumption that the human target and the mobile robot are connected by a virtual spring. The input velocity to a mobile robot is generated on the basis of an elastic force of a virtual spring, and this proposal absorbs the gap between the human and the mobile robot motion.

Another solution is the presumption based on the detailed analysis that human walking is included into the control. In this work, two stages of control are used: the first one performs the control parameters detection taking into account the human gait model while the second component corresponds to an inverse kinematic controller, which will be addressed in the next section.

3.1 Interaction Strategy for cHRI

The human-robot interaction model is shown in Fig. 3.1a. The variables and parameters used in the presented model are: human linear velocity (v_h), human angular velocity (ω_h), human orientation (ψ_h), robot linear velocity (v_r), robot angular velocity (ω_r) and robot orientation (ψ_r). The interaction parameters were defined as the angle φ between v_h and $\overline{RH}$ (named Human-Robot Line), the angle θ between $\overline{RH}$ and $\overline{RC}$ segments, and d, the length of $\overline{RH}$. Finally, the parameter a defines the distance between the controller reference point (R) and the robot center of rotation (C).

The control proposal is based on the inverse kinematics and the control variables are the angle φ and the distance d. The control law of this system aims to achieve a desired human-robot distance ($d = d_d$) and an φ angle that converges asymptotically to zero.

The components that affect the control variables are depicted in Fig. 3.1b. That way, the direct kinematics is shown in (3.1), where $\tilde{d}$ is the difference between the desired and measured distances.

$$\begin{pmatrix} \dot{\tilde{d}} \\ \dot{\tilde{\varphi}} \end{pmatrix} = \begin{pmatrix} \cos(\theta) & -a\sin(\theta) \\ -\frac{\sin(\theta)}{d} & -a\frac{\cos(\theta)}{d} \end{pmatrix} \overbrace{\begin{pmatrix} v_r \\ \omega_r \end{pmatrix}}^{u} + \begin{pmatrix} -v_h\cos(\varphi) \\ \omega_h + v_h\frac{\sin\varphi}{d} \end{pmatrix} \tag{3.1}$$

The inverse kinematics controller, obtained from the kinematic model presented in (3.1), is shown in (3.2) and (3.3).

$$v_r = \cos(\theta)\left[-k_d\tilde{d} + v_h\cos(\varphi)\right] - d\sin(\theta)\left[-k_\varphi\tilde{\varphi} - \omega_h - \frac{v_h}{d}\sin(\varphi)\right] \tag{3.2}$$

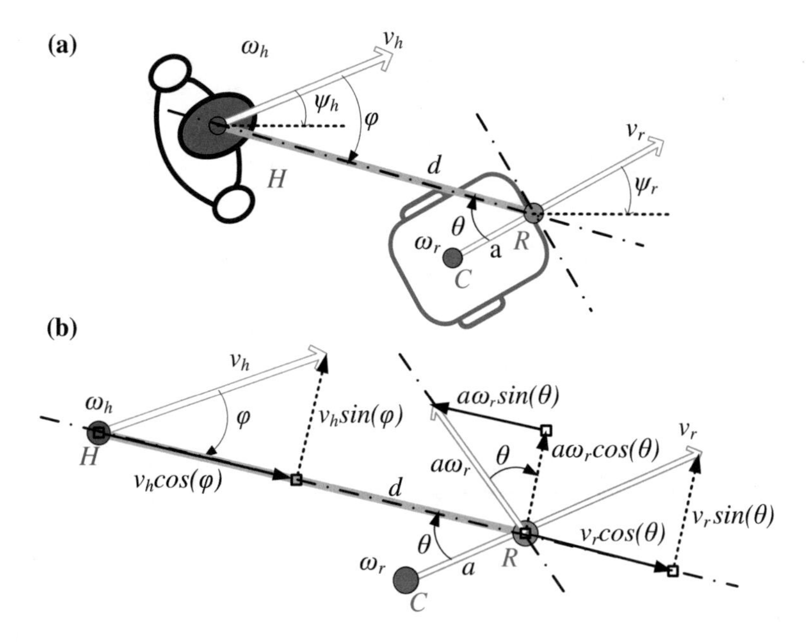

Fig. 3.1 Model for cHRI applied in a carrier robot. **a** Kinematic model. **b** Detailed kinematic model

$$\omega_r = -\frac{\sin(\theta)}{d}\left[-k_d\tilde{d} + v_h\cos(\varphi)\right] - \frac{d}{a}\cos(\theta)\left[-k_\varphi\tilde{\varphi} - \omega_h - \frac{v_h}{d}\sin(\varphi)\right] \tag{3.3}$$

In this work, no dynamics effects are assumed. This assumption is based on the fact that human gait consists of slow movements, especially in human-robot interaction scenarios, as previously observed in [10]. However, if necessary, a dynamic compensator could be integrated into the control scheme. This compensator could be obtained from an identification process [11] and used in series with the kinematic controller [12, 13]. Human dynamics are also not considered. Nevertheless, the human kinematics is here used as an input to the control law. In this context, the commands are given directly to the robot to follow the human.

In this kinematic approach, using the proposed control law and assuming a perfect velocity tracking by the robot, the control errors $\tilde{d}$ and $\tilde{\varphi}$ converge to zero. This conclusion becomes evident after substituting (3.2) and (3.3) into (3.1), thus obtaining (3.4).

$$\begin{pmatrix}\dot{\tilde{d}}\\ \dot{\tilde{\varphi}}\end{pmatrix} = \begin{pmatrix}-k_d\tilde{d}\\ -k_\varphi\tilde{\varphi}\end{pmatrix} \tag{3.4}$$

The control system is exponentially asymptotically stable, as it can be seen in (3.5) and (3.6).

$$\tilde{d}(t) = \tilde{d}(0)\,e^{-k_d t} \tag{3.5}$$

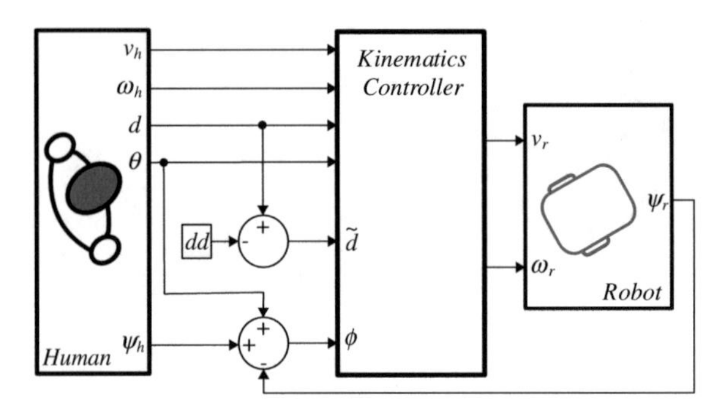

Fig. 3.2 Block diagram of the proposed controller

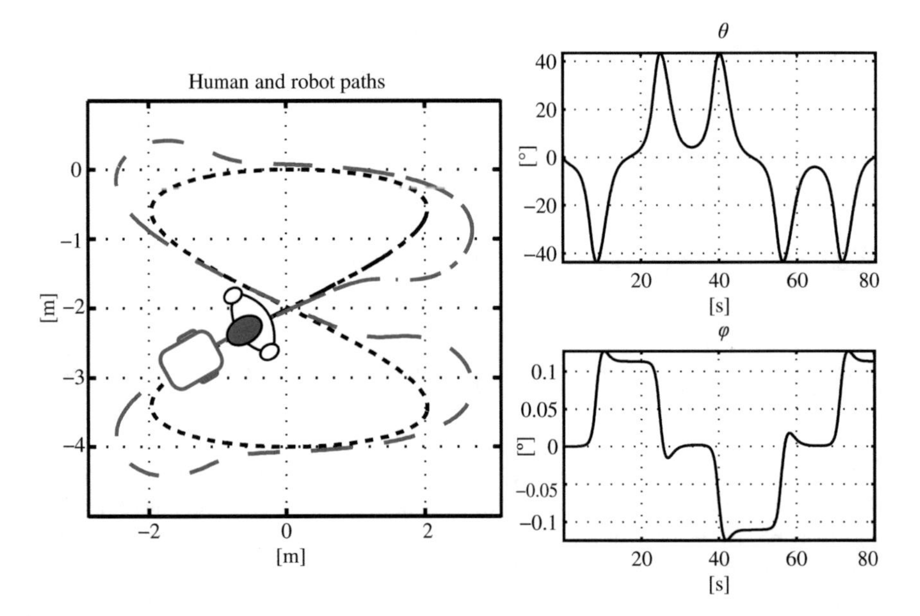

Fig. 3.3 Simulation of the proposed control strategy

$$\tilde{\varphi}(t) = \tilde{\varphi}(0)\, e^{-k_{\varphi} t} \tag{3.6}$$

The control structure here proposed is shown in Fig. 3.2, where the control errors are $\tilde{d}$ and $\tilde{\varphi}$. The error $\tilde{\varphi}$ can be obtained as a function of θ, ψ_h and ψ_r (Fig. 3.1a). The other inputs to the controller are v_h, ω_h, d and θ. The controller outputs are the control actions, such as v_r and ω_r.

The proposed control strategy was simulated with different human locomotion patterns (straight lines, circle-shaped and eight-shaped paths, etc.) in order to observe whether the walker correctly follows the user. Figure 3.3 shows one of the proposed simulations in which a human path performing an eight-shape curve (input) and the walker path following the human in front (controller output) are shown.

This simulation shows the stability of the controller even with sharp curves performed by the human. It can be observed how the θ angle is close 30° making a turn, and φ is kept less than 1° (Fig. 3.3). Therefore, the proposed controller is expected to keep the robot continuously following the human while maintaining itself positioned in front of the user.

It is possible to state that a good real-time implementation of the method proposed in this section relies on robust and precise measurement or estimation of the variables used in the control scheme (see Eqs. 3.2 and 3.3). Consequently, control inputs estimation has a paramount importance in this approach. The next sections describe and validate a method to obtain the control parameters.

3.2 Estimation of Control Inputs

The estimation of the control inputs (see Fig. 3.2) is described in this section, which is organized as follows. Firstly, the approach description based on the sensor combination of both LRF and IMU sensor is shown. Secondly, the robot and sensor system setup are also shown. Finally, the algorithm to estimate such parameters based on actual signals from the sensor setup is presented.

In this approach, human walking information and spatio-temporal gait parameters are included into the strategy for the estimation of the interaction parameters. Indeed, control inputs (set-points) are updated at each gait cycle. At the end of each gait cycle, controller outputs are calculated and sent to the robot. At the same time, the new parameter detection process starts with the next gait cycle. This parameters detection process will be explained in the following section.

The gait cycle is divided into two phases: stance and swing. Both the beginning and the end of stance involve a period of bilateral foot contact with the floor (double support). Alternatively, during the swing phase, the foot is in the air and the leg is swinging through preparation for the next foot strike [14] (Fig. 3.4). In this approach, legs position and orientation are obtained from the LRF, which is located on the robot. This information enables the estimation of parameters related to legs' kinematics and human position from the robot.

Moreover, the hip represents the junction between the passenger and the motor units. It provides three-dimensional motion with specific muscle control for each direction of activity. During the stance phase, the primary role of the hip muscles is stabilization of the superimposed trunk. In the swing phase, limb control is the objective. During each stride, the pelvis moves asynchronously in all three directions. The site of action is the supporting of the hip joint. Consequently, the greatest amount of motion occurs at the pelvis.

All motions follow small arcs, representing a continuum of postural change [15]. The transverse plane of pelvic rotation is also shown in Fig. 3.4 (segmented-line) [16]. Consequently, the pelvis allows to capture important gait kinematics information regarding the oscillatory components that are included into the human gait. These oscillations represent the main sources to detect the control parameters in this

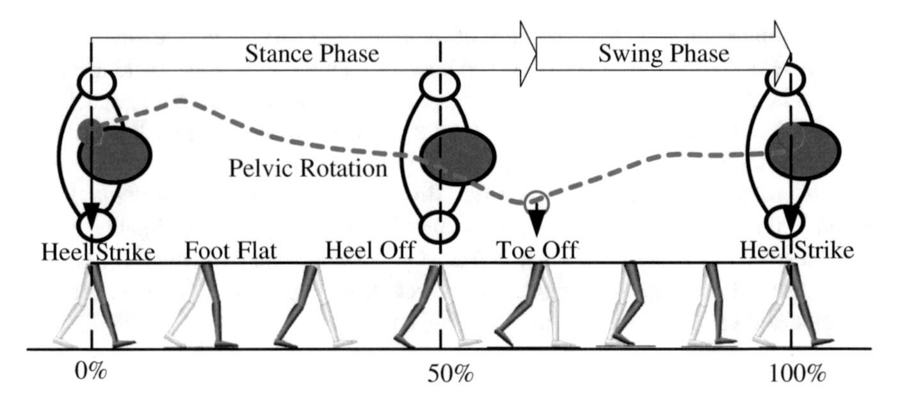

Fig. 3.4 Gait phases and pelvic rotation (transverse plane)

approach. That way, human orientation and human angular velocity estimation are obtained by an IMU sensor located on the human pelvis.

An example of the pelvis motion during a normal gait cycle on the transversal plane is depicted in Fig. 3.5 (dashed line). The method to obtain the parameters of the proposed model is as follows:

1. Human linear velocity (v_h) is the rate of change of the position in each stride. Therefore, during the human walking, it is necessary to detect the beginning and the end of the gait cycle.
2. Human angular velocity (ω_h) is the average value of the angular velocity during each cycle gait. This velocity is measured in this approach from the rate change of the pelvic rotation.
3. Human orientation (ψ_h) is the average value of the pelvic rotation during each cycle gait.

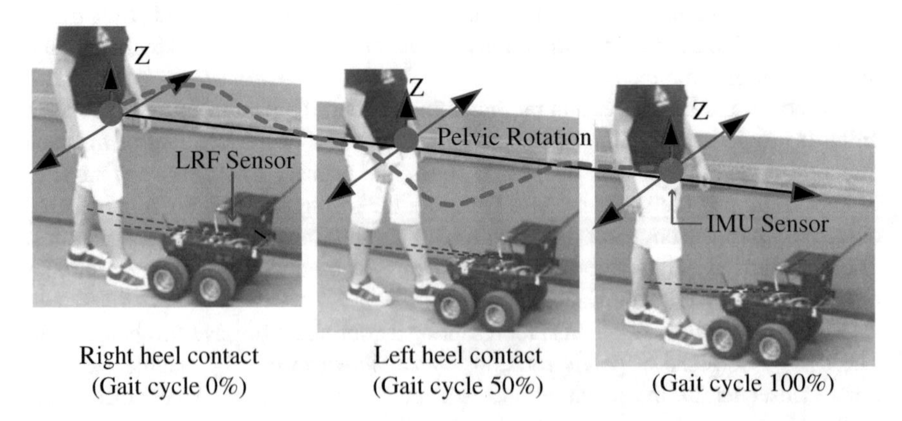

Fig. 3.5 External and internal gait measurements when the robot is following the human in front

4. Robot orientation (ψ_r) is measured by the robot odometry sensors. However, an onboard IMU sensor can be used in order to get a more accurate measurement.
5. θ represents the human orientation in relation to the robot. In order to get an accurate measurement despite the human is walking, θ should be measured when both legs have equal distance from the robot (d, obtained with the LRF sensor), and at the same time, the pelvic rotation is close to zero (Fig. 3.5).
6. φ represents the difference in orientation between v_h orientation vector and the human-robot segment $\overline{RH}$ (Fig. 3.1a). φ is also equal to $\theta - \psi_r + \psi_h$ (Fig. 3.2). This angle is only defined if the magnitude of the v_h is greater than zero.

3.2.1 Robot and Sensor System Setup

A mobile robot Pioneer 3-AT [17] was used for the practical validation of the interaction scheme presented in the previous sections. The robot has an onboard computer with a Wi-Fi link, which receives the robot state as well as the control information, such as angular and linear velocities, as shown in Fig. 3.6.

The maximum linear velocity is set to 0.7 m/s and the maximum angular velocity is set to 140°/s. It can also be seen in Fig. 3.6 a SICK LMS-200 LRF [18], which is mounted at the leg's height level, with an angular resolution of 1°.

The IMU sensor used to measure the human pelvic motion was developed in a previous research [19, 20], which is a wearable ZigBee IMU called ZIMUED. This sensor node is capable of sending data such as 3D accelerations, 3D angular velocities, 3D magnetic information and orientation information (roll, pith and yaw) through ZigBee to the ZIMUED Coordinator. This sensor is attached to the human pelvis as shown in Fig. 3.7.

The robot and sensor system integration setup has two possible configurations as can be seen in Fig. 3.6. The first one is the evaluation of the human-interaction parameters, where a remote computer receives the LRF data and robot orientation through WI-FI link. In this mode, the controller is not executed, but it is useful to

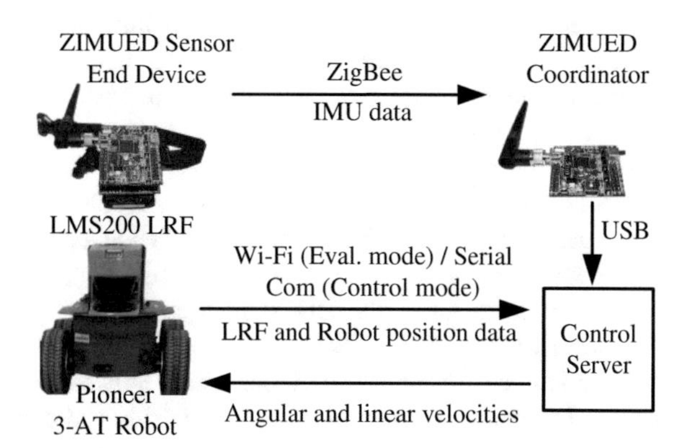

Fig. 3.6 Robot and sensor integration setup

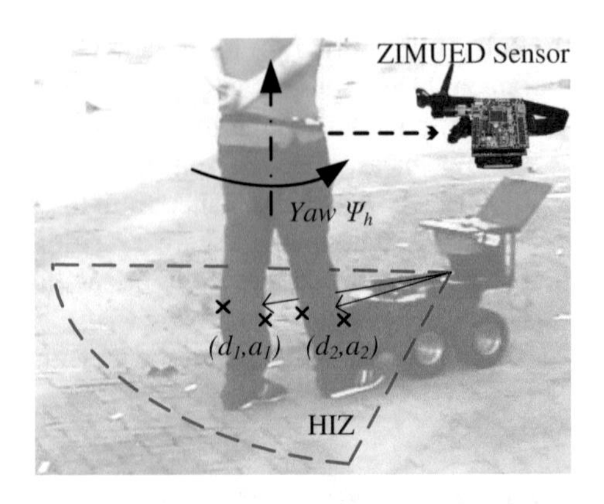

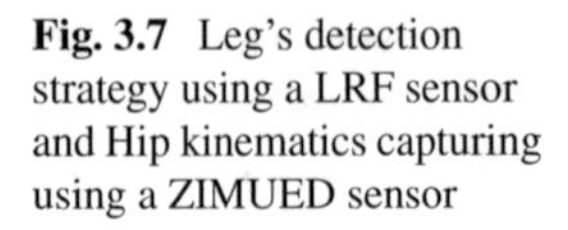
Fig. 3.7 Leg's detection strategy using a LRF sensor and Hip kinematics capturing using a ZIMUED sensor

analyze the performance of the parameters detection algorithm. The second mode is the control mode, in which the onboard computer receives the sensor information to execute the controller. A ZIMUED coordinator is linked by ZigBee connection with an IMU sensor on the human. In the same way, the coordinator sends the human IMU data to a USB connection with the computer for both configurations. LRF and robot states are sampled every 100 ms and the ZIMUED sensor at every 20 ms. At the same time, the robot is able to receive the control commands such as angular and linear velocities to be performed.

In the control mode, the main program receives IMU data every 20 ms. This packet defines the main clock of the detection algorithm. The performance of the communication setup was evaluated in [19].

Considering the sample and transmission conditions used in this setup, the wireless communication does not present problems regarding lost data packages. However, if the controller is executing and suddenly the ZigBee communication link is broken, the detection algorithm is blocked, and an internal timer is started. If no packets are received within 100 ms, the robot is automatically stopped, guaranteeing a safe operation.

The leg detection approach presented in this work combines techniques presented in [21, 22], which is split into four basic tasks: LRF data pre-processing, transitions detections, pattern's extraction and estimation of legs' coordinates. In the pre-processing phase, the delimitation of the HIZ (Human Interaction Zone) is performed (Fig. 3.7), and then laser scanning data are used to identify transitions.

The legs' positions are calculated in polar coordinates (Fig. 3.7). The general process is based on the differences between two transition events that define a leg pattern (x-marks on Fig. 3.7). After that, both distance and angle measurements are calculated in relation to the middle point of each leg. In Fig. 3.7, (d_1, a_1) and (d_2, a_2) represent the polar coordinates of left and right legs, respectively.

The angle range of the HIZ is restricted from $-60°$ to $60°$, and the scanning distance from the LRF is limited up to 2 m. On this range, the human can walk freely but the legs cannot present any occlusion. When one leg cannot be detected as a cause of screening by the other leg, the algorithm calculates the human distance with the only one leg detected. Finally, in the case the human leaves the HIZ, the robot is automatically stopped.

3.2.2 Estimation of Interaction Parameters

The parameter estimation here proposed is based on the leg detection from the LRF and pelvic rotations (see Fig. 3.5) obtained from the IMU sensor (Fig. 3.6). This signal is represented by the yaw orientation. The velocity of this orientation is periodical due to the cyclic nature of human gait, making this signal suitable to synchronize the parameter estimation every gait cycle. In Fig. 3.8, the signals of the pelvic motion and laser detection of the Right and Left Legs (RL and LL) distances are shown. These measurements were obtained through experiments with a person walking towards the LRF sensor.

Figure 3.8a shows the pelvic angular velocity obtained from the Z axis gyroscope signal. The zero crossing points are marked with a circle and square at every gait cycle. Figure 3.8b shows the square mark representing the maximum pelvic orientation

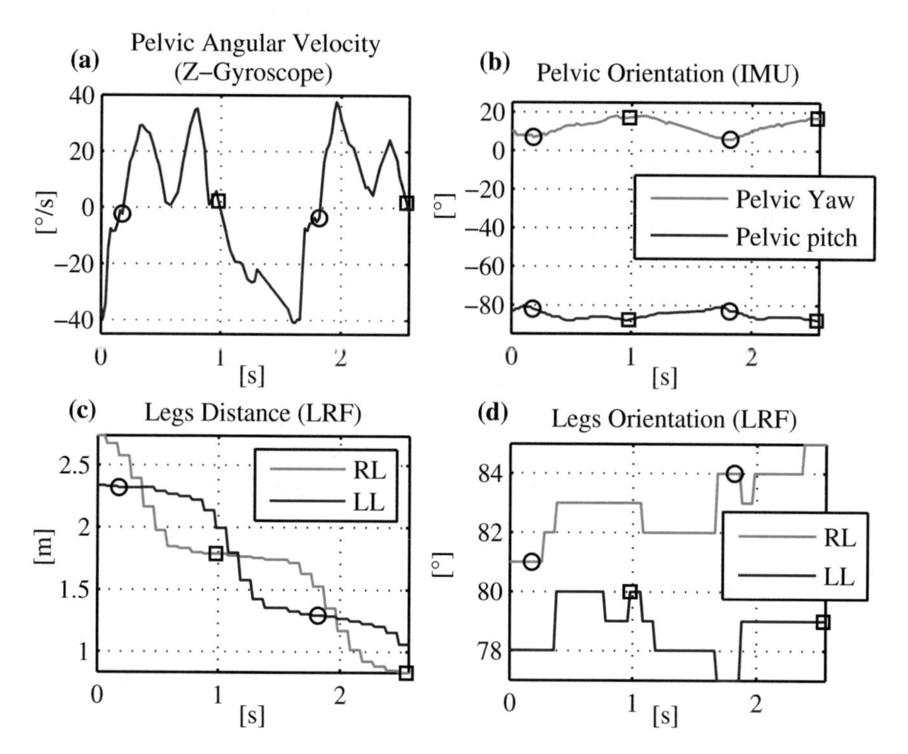

Fig. 3.8 Detection of zero crossing points over pelvic angular velocity. **a** Pelvic Angular Velocity (Z-Gyroscope). **b** Pelvic Orientation (IMU). **c** Legs Distance (LRF). **d** Legs Orientation (LRF)

(they happen after the right heel contacts the ground). The circle mark represents the minimum pelvic orientation (it happens after the left heel contact). At the same time, these events are presented in the RL and LL distances and orientation trajectories respectively (Fig. 3.8c, d). The parameter detection methodology is performed as follows:

1. Human linear velocity (v_h). This parameter is updated at each step. The interval between the last two zero crossing points represents the step time. The step length is the distance performed in one step. It is obtained from the maximum distance between right and left legs during the step time. The magnitude of v_h is the step length divided by the step time. Due to the fact that the robot linear velocity is limited to 0.7 m/s, during interaction experiments, the user is instructed not to exceed this limit.
2. Human angular velocity (ω_h). This parameter is calculated at each stride. It is the average of all values of angular velocity (from Z-Gyroscope) during one stride. Therefore, if the human is walking straight, the oscillatory form of the gait ω_h will be close to zero (see Fig. 3.6). Although the robot angular velocity is limited to 140°/s, this does not cause any problem as the human does not achieve such high angular speed during normal interaction with the robot.
3. Human orientation (ψ_h). This parameter is calculated at each stride by averaging all values of the pelvic orientation (from pelvic yaw) during one stride. The range of this angle is between $-180°$ and $180°$.
4. Robot orientation (ψ_r). The orientation is measured by the robot odometry at each step. The range of this angle is between $-180°$ and $180°$. Despite the odometry is the most widely used method to obtain the robot position, there are well known errors from this measurement method [23]. A more accurate measurement could be obtained by using an IMU mounted on the robot. The use of an IMU is especially important during experiments that last several minutes, as the cumulative odometry errors are more significant.
5. θ angle and human-robot distance (d). θ is the average between right and left legs orientation from the LRF legs detection. The range of this angle is restricted between $-60°$ and $60°$. This is calculated when both legs have the same distance (crossing point); thus, the human-robot distance is obtained. For interaction purposes, this distance is limited to a maximum of 2 m.
6. φ Angle. This angle is calculated as $\theta - \psi_r + \psi_h$ at each stride.

3.3 Experimental Study

Three different preliminary experiments were developed in order to verify the accuracy in the detection of the human-robot interaction parameters with the proposed algorithm. In the first and second experiments, no motion was performed by the robot. The subject was asked to walk on a straight line following different paths marked on the floor to define specific angular parameters (θ, φ and ω_h). The parameter v_h

was defined during each test according to the human gait and compared with the estimated velocity.

In the third experiment, the robot is configured with specific linear (v_r) and angular (ω_r) velocities, and the human follows the robot keeping a constant distance. Human linear and angular velocities are estimated in a more dynamic scenario and are compared to the reference velocities performed by the robot.

The layout of the paths for the first experiment is shown in Fig. 3.9a. These paths, marked on the floor (black dashed lines), have different predefined θ angles with respect to the LRF reference: $-20°$, $-15°$, $-10°$, $-5°$, $0°$, $5°$, $10°$, $15°$ and $20°$. A volunteer was asked to walk on a straight line in the direction of the robot, performing three repetitions of each one of the proposed paths. The assumption was that both θ measured from LRF and ψ_h measured from the IMU should have the same value to the predefined angles during every path, as it can be observed in Fig. 3.9b. In this experiment φ angle is always equal to zero.

The layout of the paths proposed on the second experiment is shown in Fig. 3.10a. These paths marked on the floor (black dashed lines) are performed to evaluate the φ angle estimation based on the direct measurement of θ by the LRF. Thereby, despite the fact that the start points were the same of the first experiment, all paths are now parallel to each other. The volunteer was asked to perform three repetitions of the proposed paths. Then, every path is performed by the volunteer with predetermined linear velocity (v_h) orientation, as it can be observed in Fig. 3.10b. Therefore, the assumption in this experiment is that both θ and φ have the same magnitude and opposite signs. Each path was labeled (T1, T2, T3, T4, 0°, T5, T6, T7 and T8) as shown in Fig. 3.10b.

Additionally, in the first and second experiments, each test was performed with three predefined linear human velocities (v_h): 0.25, 0.5 and 0.75 m/s to assess the effect of different gait speeds on the parameter estimation process. The selection of these velocities is based on past experience in human-robot interaction scenarios, such as carrying loads or in walker-assisted gait [10]. Thus, every path was marked with distance intervals (0.25, 0.5 and 0.75 m). In order to achieve the desired velocities, steps were performed following a sound indication produced at every second.

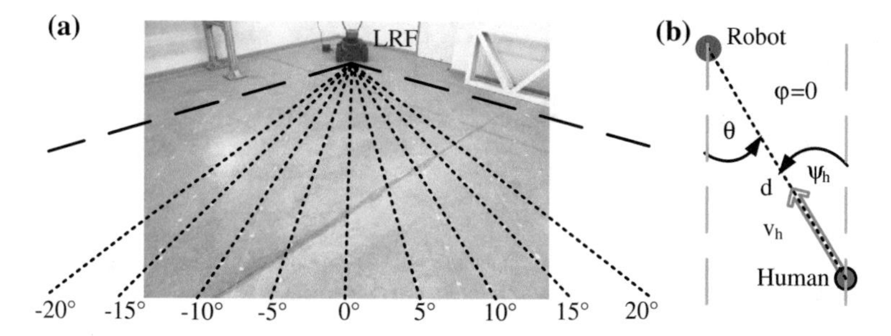

Fig. 3.9 First experiment for validation of the HRI parameters detection. **a** Proposed paths. **b** Interaction parameters

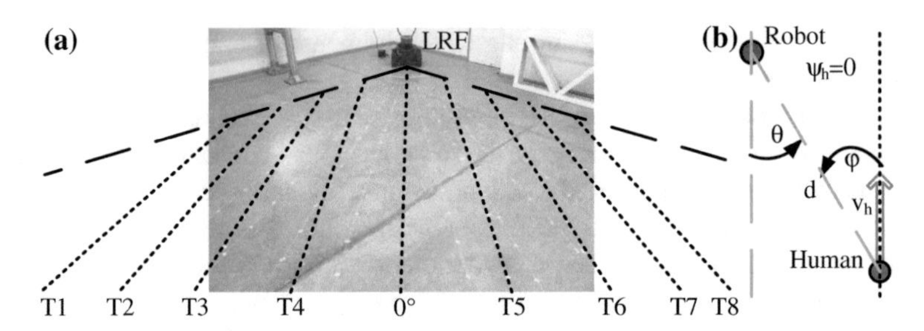

Fig. 3.10 Second experiment for validation of the HRI parameters detection. **a** Proposed paths. **b** Interaction parameters

In the first and second experiments the human angular velocity is not evaluated. Therefore, to verify the estimation process of this parameter, a third experiment was performed with a circle-shaped path (Fig. 3.11). Thus, the robot was programmed to perform constant linear and angular velocities. The human was asked to maintain a constant distance while following the robot. To simplify this task, human hands were kept in contact with the robot as shown in Fig. 3.11a. The assumption in this experiment is that human angular and linear velocities will be approximately equal to the robot's velocities (Fig. 3.11b). Three circle-shaped trajectories with different constant linear and angular velocities were programmed: (i) 0.15 m/s and −7°/s; (ii) 0.25 m/s and −11°/s; and (iii) 0.30 m/s and −14°/s.

The results of the three experiments show the precision and variability of the human-interaction parameters estimation. The first section presents the results of the experimental validation of the proposed methodology for the estimation of interaction parameters.

Once the procedure for the estimation of the interaction parameters is validated, the results of the experiments with the proposed controller are presented in the next section, showing the human-interaction parameter detection and the controller being executed, both in real-time, by the mobile robot.

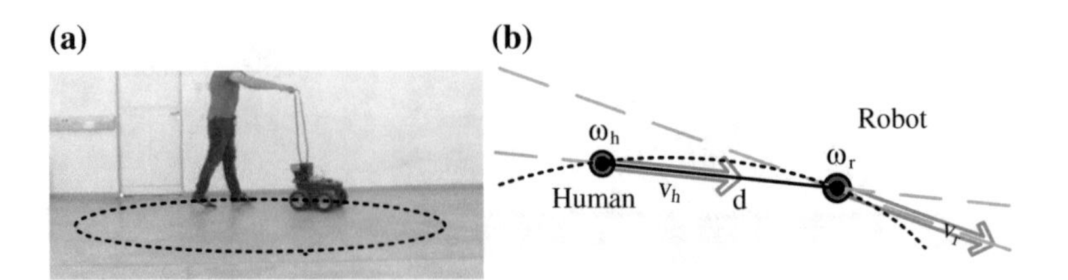

Fig. 3.11 Third experiment for validation of the HRI parameters detection. **a** Paths layout and human location to perform the circle path. **b** Interaction parameters

3.3.1 Detection and Estimation of Human-Robot Interaction Parameters

In the first experiment, θ and ψ_h estimation remain close to the expected angle in every test. Figure 3.12 shows a part of the measurements and estimated parameters performed in three predefined velocities (v1 $= 0.75$ m/s, v2 $= 0.50$ m/s and v3 $=$ 0.25 m/s) in the $-5°$ path. IMU and LRF data (continuous signals) are presented along with the human linear velocities and angular parameters (discrete values) obtained in two foot strikes.

The angular velocities obtained from the gyroscope in the z-coordinate are shown in Fig. 3.12a. As expected, there is an increase in pelvic rotation for greater linear velocities. The average of the angular velocity remains close to zero because the human is walking in a straight line. Pelvic yaw and pitch angles are shown in Fig. 3.12b, where ψ_h is obtained from the yaw angle. It is also observed an increase in the oscillation amplitude with the increase of v_h.

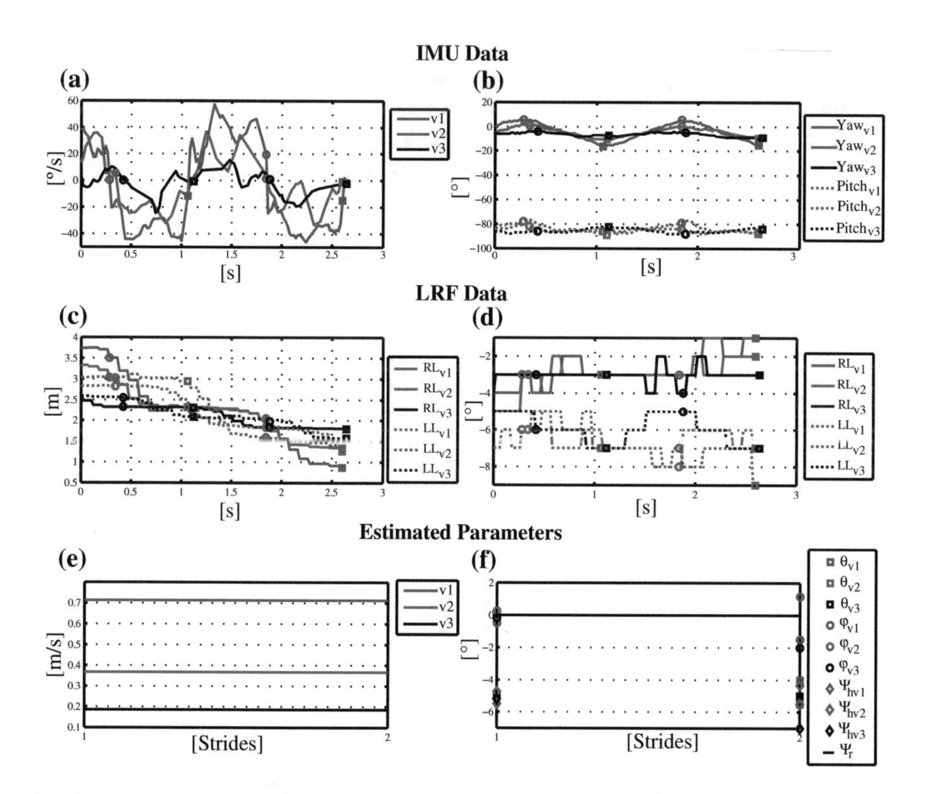

Fig. 3.12 Measurements and estimated parameters performed in the first experiment (test of $-5°$). **a** Pelvic angular velocities from z-axis gyroscope. **b** Pelvic orientation from IMU. **c** Legs' distance curves from LRF detection. **d** Legs' orientation curves from LRF detection. **e** Estimated v_h. **f** Estimated θ, φ and ψ_r angles

The paths of the human legs obtained in these intervals are shown in Fig. 3.12c. As expected, stride length increases when v_h increases. As the robot is not moving, the module of the slope of these curves is the actual v_h. The negative values of the slope indicate the decrease in the distance as the subject is walking towards the LRF. Although feet position (indication of the step length) were marked on the floor, the resolution of the step length measurements is affected by the shoe size, which is reflected on the error of the v_h estimation as shown in Fig. 3.12e.

The legs orientation obtained from the LRF detection is shown in Fig. 3.12d. Finally, the estimated angular parameters are shown in Fig. 3.12f. Note that θ and ψ_h angles were close to the expected $-5°$. Also, the φ angle is close to zero as proposed in this experiment.

From the first experiment, all estimated values of θ and ψ_h for different v_h were grouped and compared with the path angle (reference value). In the estimations of θ (Fig. 3.13a), the RMSE was 0.6° and the bias was $-0.6°$. The values obtained for the errors seem to remain constant in all experiments. This could be caused by a misalignment of the LFR sensor during the experimental setup. Regarding the estimations of ψ_h (Fig. 3.13b), the RMSE was 0.2° and the bias was $-0.2°$. Despite of the continuous oscillation of the pelvis during walking, estimation was precise and unbiased, showing also repeatability with changes of v_h.

Considering the second experiment, Fig. 3.14 shows the angular parameters during different tests in a single stride. It is possible to see that ψ_h remains close to zero, and θ and φ remain close to a same magnitude with opposite signs, as expected.

From the first and second experiments, v_h average errors (RMSE) of all tests were grouped in Fig. 3.15a. The estimation of the error for 0.25, 0.50 and 0.75 m/s remains under 0.15 m/s. Although this is high in comparison with the desired/performed

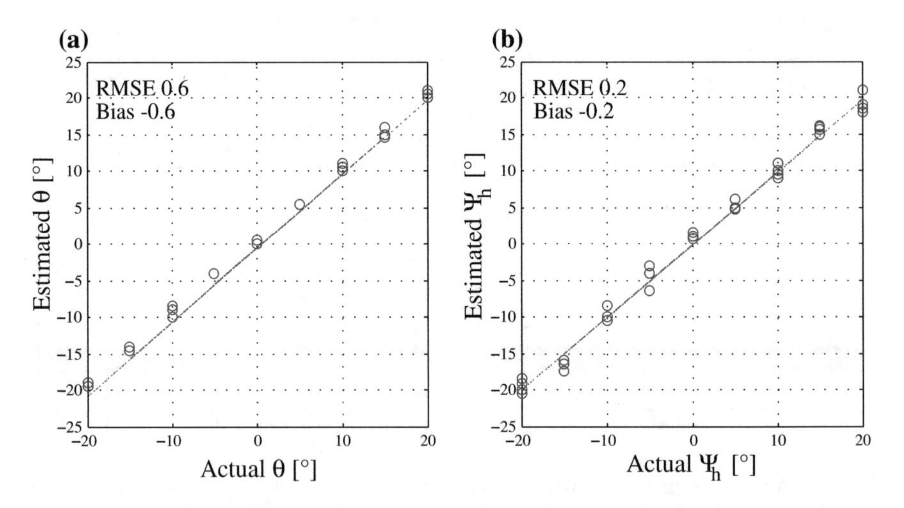

Fig. 3.13 Estimated values of θ and ψ_h versus reference angles from the first experiment. **a** Estimated θ in first experiment. **b** Estimated ψ_h in first experiment

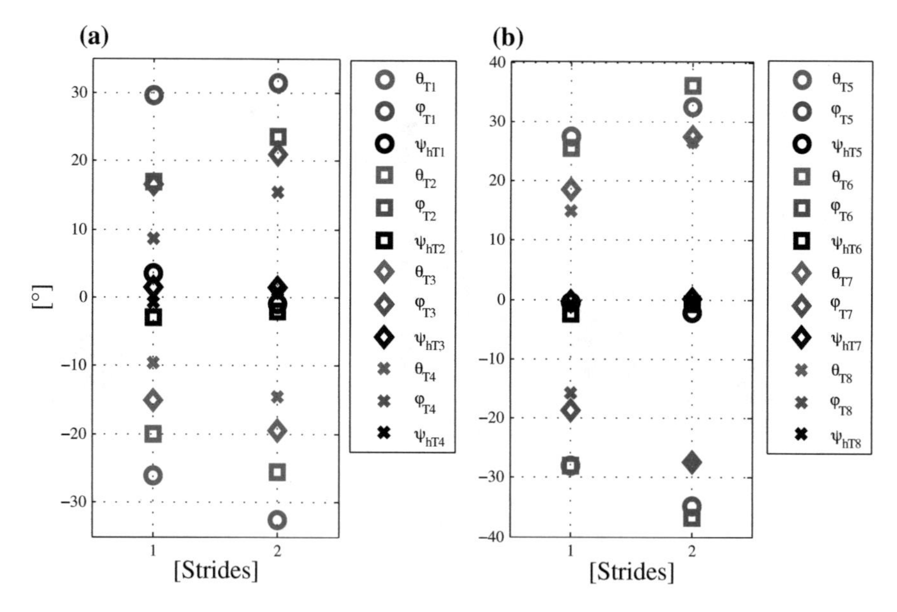

Fig. 3.14 Estimated values of θ, φ and ψ_h from the second experiment. **a** Tests T1 T2 T3 T4 HLV 0.5 m/s. **b** Tests T5 T6 T7 T8 HLV 0.5 m/s

speed, it is important to mention that errors may be caused by a misplacement of the feet in two consecutive steps. To illustrate this, one could imagine the situation in which the human steps the line with the toe on a step and with the heel on the consecutive one. Considering that the foot size presents magnitudes in the same order as the step lengths, errors with the presented magnitude are expected in these experiments.

Additionally, the errors of the angular parameters (Fig. 3.15b) remain close to 3°. The error of θ is considerably smaller (around 1°) due to the direct measurement of this parameter using the LRF, which presents higher resolution.

In the third experiment, the robot follows constant angular and linear velocities describing a circle-shaped path. Figure 3.16 shows a part of the measurements and the estimated parameters for the robot trajectory performed for linear velocity of 0.3 m/s and angular velocity of −14°/s.

The angular velocities obtained from the gyroscope in the z-coordinate are shown in Fig. 3.16a. Due to the performed circle path, the estimated ω_h remains close to −14°/s as expected (Fig. 3.16f). This measurement can also be observed in the tendency of the pelvic orientation values shown in Fig. 3.16b.

The position and orientation of the human legs obtained in this interval are shown in Fig. 3.16c, d, respectively. Due to the fact that the LRF and the legs are moving at the same time, it can be observed that these signals present a constant mean value. The v_h estimation is shown in Fig. 3.16e and remains close to the expected 0.3 m/s. During the tests, the human was following the robot. This can be observed through

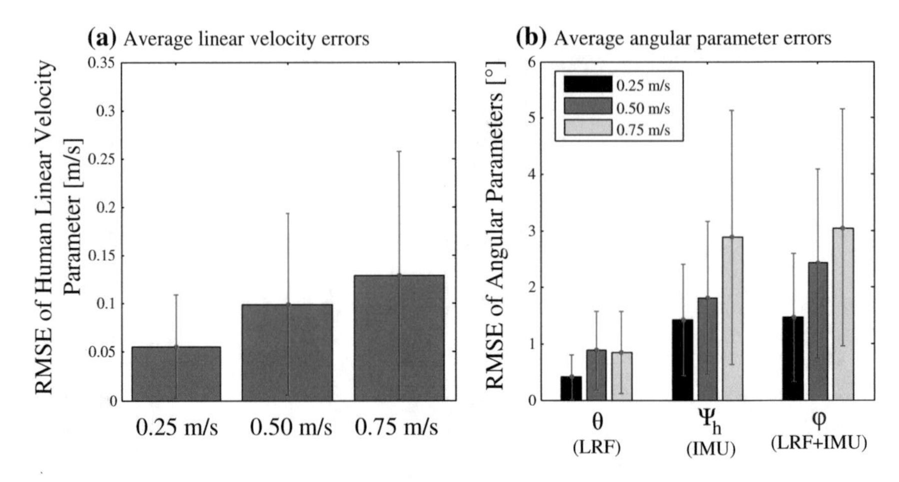

Fig. 3.15 Average errors (RMSE value) in estimation of v_h for 0.25, 0.50 and 0.75 m/s. **a** Linear velocity estimated errors. **b** θ, ψ_h and φ, and errors with the different velocities

the pattern of the ψ_h and ψ_r angles in Fig. 3.16g. As a result of that, θ and φ are shown in Fig. 3.16h.

Table 3.1 shows the summary of the actual and estimated linear and angular velocities in the third experiment. The linear velocity error corresponds to the previous analysis, and the angular velocity error remains close 1°/s, which is acceptable in this kind of interaction strategy.

3.3.2 *Controller Evaluation*

After the validation of the parameter estimation methodology, a final experiment with the robot following in front of the user was conducted. In this experiment, a volunteer performed the eight-shaped path (lemniscape) shown in Fig. 3.17. During the execution of turns the robot follows the humans on the external side when he/she is making a curve (Fig. 3.3a). The human path and the expected robot's path (solid line) can be observed in Fig. 3.17 that also shows the start and the end marks of the human path; the human walks in a straight line before entering the eight-shaped path. It is noteworthy that the eight-shaped curve is analyzed in three phases: first, a semicircle path (human turning left); second, a circle path (human turning right); and third, a last semicircle path (human turning left). This way, it is possible to analyze the performance of the controller in straight and in curve-shaped trajectories.

Figure 3.18 shows the IMU and LRF sensor data obtained during the proposed experiment. Although there are periodic (and tridimensional) oscillations of the pelvis during the gait and considering that the locomotion was performed in an

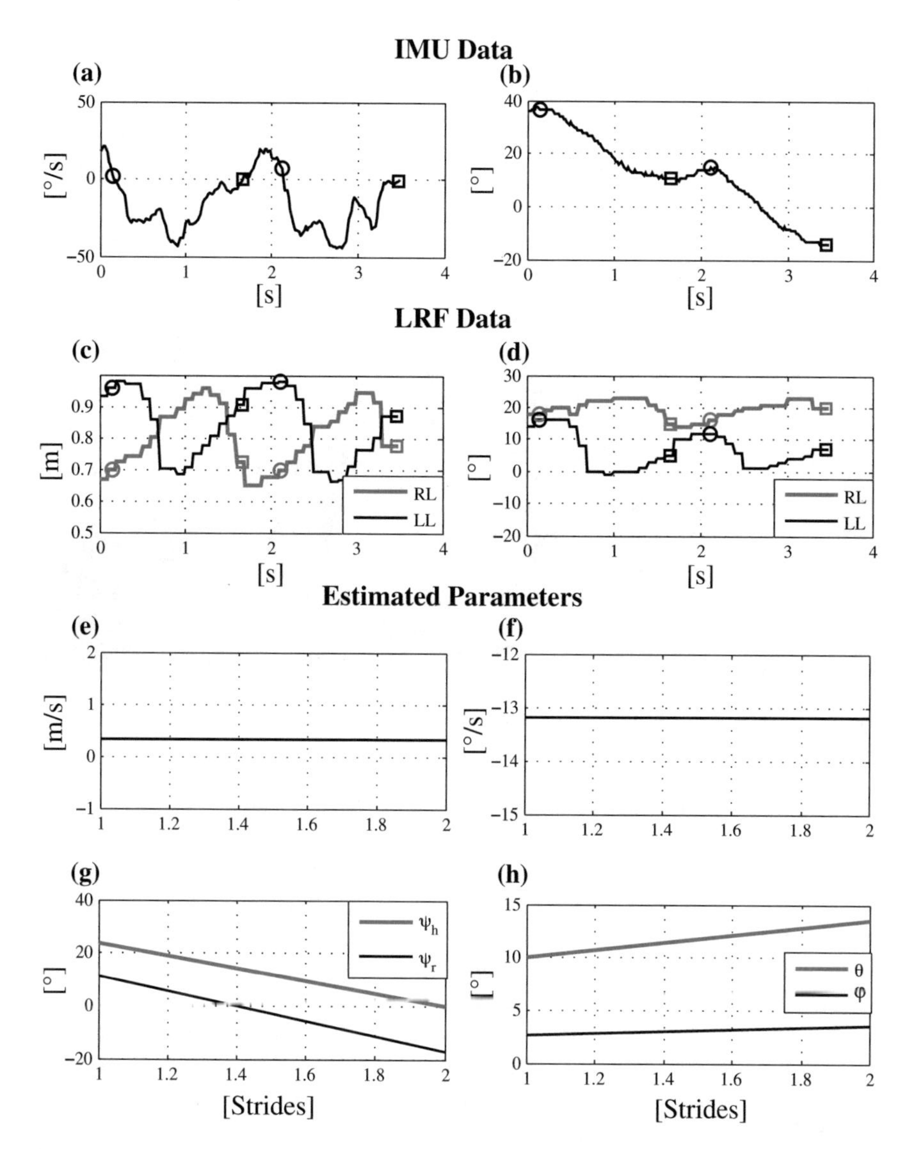

Fig. 3.16 Measurements and estimated parameters with $v_h = 0.3$ m/s and $\omega_h = -14°$/s (third experiment). **a** Pelvic angular velocity from z-axis gyroscope. **b** Pelvic orientation from the IMU. **c** Legs distance *curves* from LRF detection. **d** Legs orientation curves from LRF detection. **e** Estimated v_h. **f** Estimated ω_h. **g** Estimated ψ_h and ψ_r. **h** Estimated θ and φ angles

Table 3.1 Error in estimation of linear and angular velocities for the third experiment

Actual v_h (m/s)	Actual ω_h (°/s)	Estimated v_h (m/s)	Estimated ω_h (°/s)	Error v_h (%)	Error ω_h (%)
0.15	−7	0.149	−6.6	1	5
0.25	−11	0.253	−10.4	−1	5
0.3	−14	0.311	−13.2	−4	6

Fig. 3.17 Human path (*dashed line*) performing an eight-shaped curve (lemniscape)

eight-shaped path, the robot kept a continuous and stable orientation while following, as shown by ψ_r (gray line) in Fig. 3.18a.

Figure 3.18b shows the raw signal obtained from the gyroscope placed on the human pelvis (gray line) and the filtered signal (black line). A second order Butterworth low-pass filter (cutoff frequency of 1 Hz) was used to reject high frequency components that are not associated with the gait cadence. As it can be seen, no significant delay was observed in this application.

The legs detection was adequate during the whole experiment as depicted in Fig. 3.18c (angle detection), d (distance detection). The values of angular positions of the legs, measured from the robot, were in the range between −40° and 40° (Fig. 3.18c). These bounds belong to the range of detection previously defined [−60°, 60°]. In this experiment, the maximum interaction distance was set to 2 m and the desired distance d_d was set to 0.9 m. Accordingly, the legs distance measurements were between 0.4 m and 1.2 m during the whole the experiment (Fig. 3.18d).

Figure 3.19 shows snapshots of different instants of the experiment illustrated in Fig. 3.17, which lasted about 80 s. From the beginning and up to the fifteenth second, the human walked in a straight line (Fig. 3.19a). After that, the human began to turn left (ψ_h in Fig. 3.18a) entering the eight-shaped path. The first semicircle is performed up to about the 30th s (Fig. 3.19b). The human orientation increased positively in this interval (Fig. 3.18a), indicating that he was turning left. The orientation of the legs

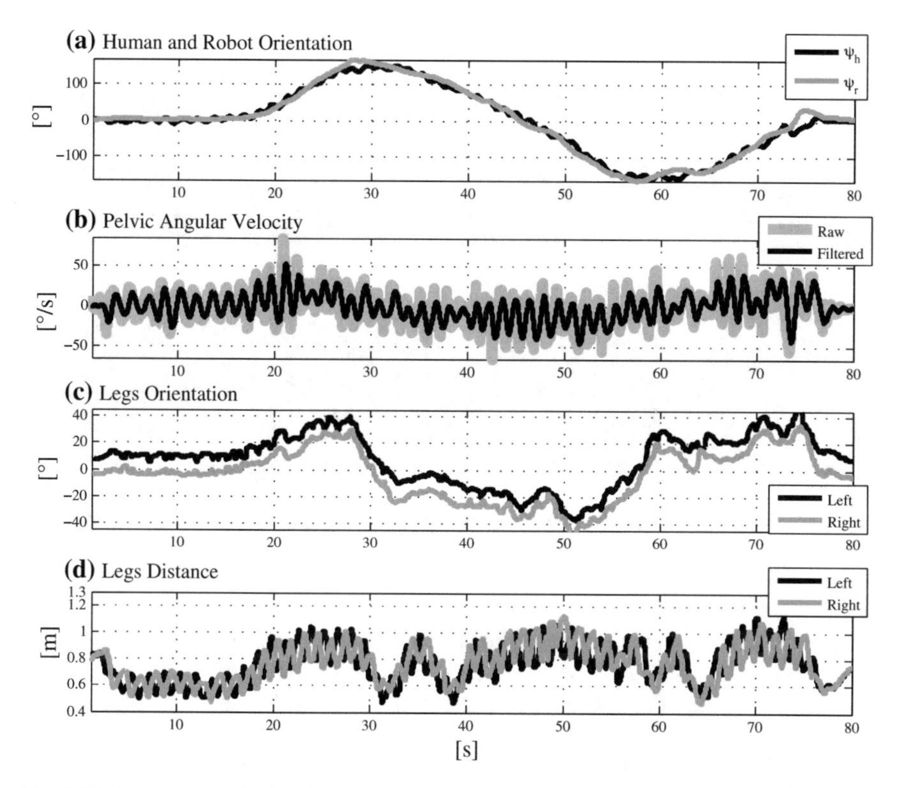

Fig. 3.18 Sensors data of robot following in front of the user performing an eight-shaped curve. **a** Human and robot orientation from IMU and robot odometry respectively. **b** Pelvic angular velocity from gyroscope (raw data and filtered signal). **c** Leg orientation measured with the LRF sensor. **d** Leg distance measured with the LRF sensor

(LRF data) decreased to 0° (Fig. 3.18c) before finishing the first semicircle as the human starts planning the next circle (Fig. 3.19c).

This circle is completed before the 60th s (Fig. 3.19e). In this interval, the human orientation decreases constantly, as expected (see Fig. 3.18a), indicating that he is turning right. After that, the angular positions of the leg become 0° again (Fig. 3.18c) in order to perform the last semicircle (Fig. 3.19e).

Finally, the human is back at the beginning of the eight-shaped curve (Fig. 3.19f). ψ_h and ψ_r angles are close to 0° again, as expected (Fig. 3.18a).

As aforementioned, all control parameters are detected every gait cycle. Some of them are updated every step while others are updated at every stride. However, the controller variable update is executed at every step. In the case that human does not perform another step, for example, when the human suddenly stops, the parameters are calculated at every second. Finally, Fig. 3.20 shows all control data recorded during the proposed experiment. The parameters estimation algorithm detects approximately 100 steps from the human in the execution of the proposed path.

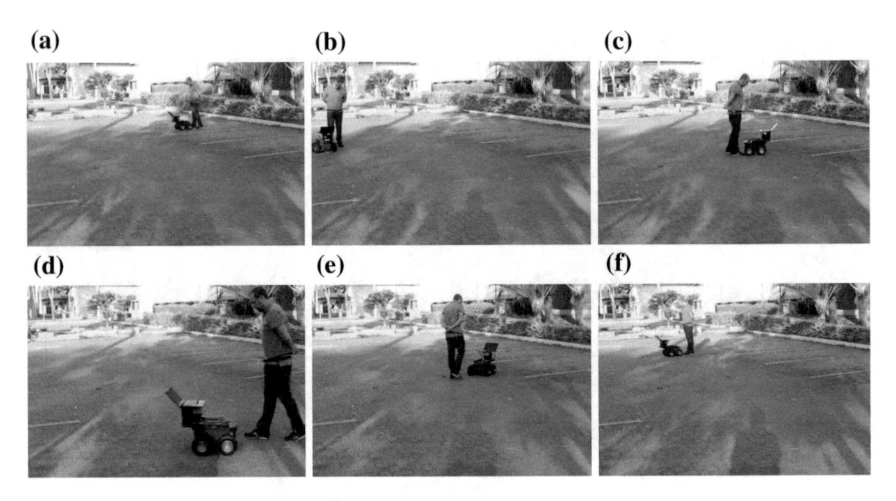

Fig. 3.19 Snapshots of the experiment performing an eight-shaped curve by the user, where the robot is following in front the user

In Fig. 3.20a, from the beginning and up to almost the step number 20, the human was walking in a straight line, as ψ_h, ψ_r and θ remains close to 0° (Fig. 3.20a). This way, φ (control error) remains close to 0°, as well. However, $\tilde{d}$ remains near −0.3 m (Fig. 3.20b). As a result of this, the control action $v_r(C)$ and the robot's actual speed $v_r(R)$ follow v_h with a maximum value of approximately 0.3 m/s (Fig. 3.20c). Furthermore, the control action $\omega_r(C)$ and the measured velocity $\omega_r(R)$ remain close to 0°/s (Fig. 3.20d), as expected.

After the step number 20, the eight-shaped curve starts. From Fig. 3.20a ψ_r follows ψ_h continuously, θ is positive when the human is turning left and negative when the human is turning right, and remains close to 0°, as expected.

From Fig. 3.20b, $\tilde{d}$ was negative in most of the experiment. This indicates that the human walks forward and the controller tries to reach the desired distance (0.9 m). From Fig. 3.20c, v_h was always lower than 0.5 m/s, however the control action, $v_r(C)$, reaches the robot's maximum forward speed (0.7 m/s) and also sometimes the backward speed limit (−0.7 m/s). The controller tries to bring the control errors to 0. $v_r(R)$ is delayed with respect to $v_r(C)$ due to robot dynamics, but this delay does not significantly affect the performance of the controller response with these experiment conditions. From Fig. 3.20d, $\omega_r(C)$ and $\omega_r(R)$ present an adequate tracking of ω_h, but also there is an expected delay between $\omega_r(C)$ and $\omega_r(R)$, which is smaller than the delay between $v_r(C)$ and $v_r(R)$.

Finally, the trajectory performed during this test is shown in Fig. 3.20e. The black dashed line is the human path measured from the LRF, and the gray line represents the mobile robot path measured by the robot odometry. The triangles marks represent the starting and final points of every path.

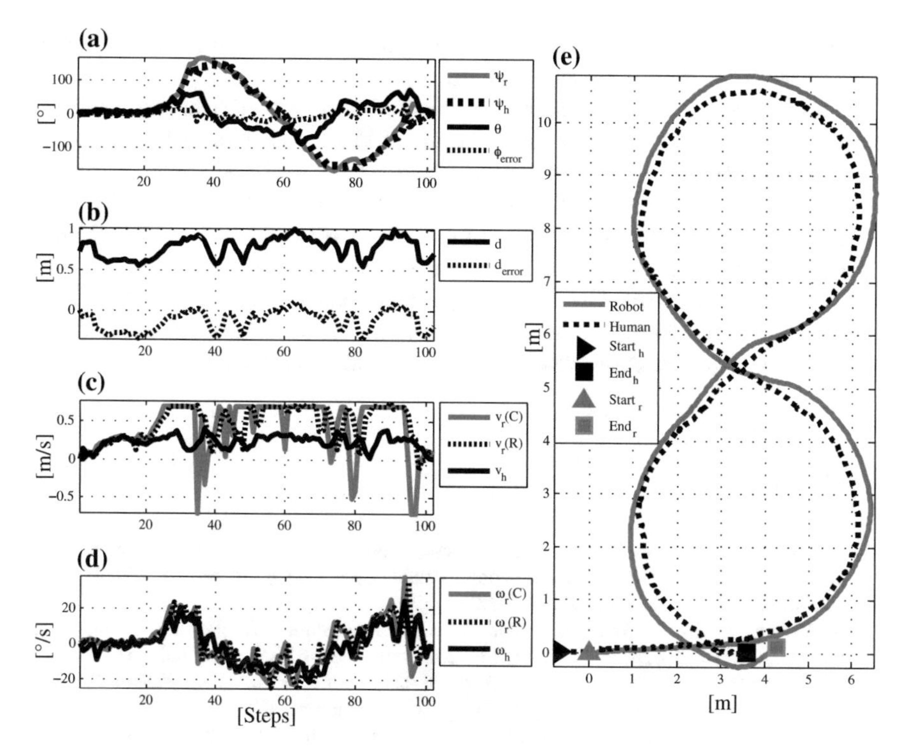

Fig. 3.20 Control data of robot following in front experiment performing an eight-shaped curve. **a** Angular parameters. **b** Distance parameters. **c** Linear velocities: control action $v_r(C)$ and measured $v_r(R)$ and v_h. **d** Angular velocities: control action $\omega_r(C)$, and measured $\omega_r(R)$ and ω_h, **e** Trajectory performed

3.4 Chapter Conclusions

This chapter presented a new human-robot interaction strategy based on the human gait by data fusion from a wearable IMU and an onboard LRF. Also, a new mobile-robot human controller for tracking in front of the human with an experimental validation of the controller performance was presented.

In the experimental study, despite of the continuous oscillation during the walking, the parameters estimation was precise and unbiased, showing also repeatability when human linear velocity changes. In the same manner, the estimation errors were lower than 10 % when the robot performed a curve-shaped path.

This research shows that the proposed control is effective in assisting a mobile robot to follow a human. A satisfactory result was obtained in terms of stable performance, through the tracking algorithms here proposed. The controller was evaluated with an eight-shaped curve (lemniscape), showing stability of the controller even with sharp changes in the human path. The controller keeps the robot continuously following in front of the human gait in all experiments. It is also shown the good performance of the controller regarding the robot orientation when it is following the human turning during the experiments.

One of the advantages of the human-interaction here proposed is the computational efficiency due to direct measurement of the human kinematics with the IMU wearable sensor on the pelvis and the legs detection from the LRF. The detection and human tracking from the mobile robot is completed in real-time and also in unstructured environments. The reliability of this approach is guaranteed with the integration of the analysis of human walking into the control parameters detection.

The next chapter will address the integration of this control strategy in a robotic walker. Some remarks regarding the human-robot physical link will demand new algorithms and validations to develop a natural walker-assisted gait based on cHRI.

References

1. C.A. Cifuentes, A. Frizera, R. Carelli, T. Bastos, Human-robot interaction based on wearable IMU sensor and laser range finder. Robot. Auton. Syst. **62**(10), 1425–1439 (2014). ISSN 09218890. doi:10.1016/j.robot.2014.06.001
2. W. Chung, H. Kim, Y. Yoo, C.-B. Moon, J. Park, The detection and following of human legs through inductive approaches for a mobile robot with a single laser range finder. IEEE Trans. Ind. Electron. **59**(8), 3156–3166 (2012). ISSN 0278-0046. doi:10.1109/TIE.2011.2170389. http://ieeexplore.ieee.org/lpdocs/epic03/wrapper.htm?arnumber=6032092
3. J. Ziegler, H. Kretzschmar, C. Stachniss, G. Grisetti, W. Burgard, Accurate human motion capture in large areas by combining IMU- and laser-based people tracking. In *IEEE/RSJ International Conference on Intelligent Robots and Systems*, (IEEE, New York, 2011), pp. 86–91 . ISBN 978-1-61284-456-5. doi:10.1109/IROS.2011.6094430. http://ieeexplore.ieee.org/lpdocs/epic03/wrapper.htm?arnumber=6094430
4. E. Prassler, D. Bank, B. Kluge, M Hagele, Key technologies in robot assistants: motion coordination between a human and a mobile robot. Trans. Control, Autom. Syst. Eng. **4**, 56–61 (2002). http://www.ijcas.org/admin/paper/files/4-1-9.pdf
5. A. Ohya, T. Munekata, Intelligent escort robot moving together with human-interaction in accompanying behavior-. In *Proceedings of the 2002 FIRA Robot Congress* (2002), pp. 31–35. ISBN 8129853515
6. E.-J. Jung, B.-J. Yi, S. Yuta, Control algorithms for a mobile robot tracking a human in front. In *Proceedings of the 2012 IEEE/RSJ International Conference on Intelligent Robots and Systems*, (IEEE, New York, 2012) pp. 2411–2416. ISBN 978-1-4673-1736-8. doi:10.1109/IROS.2012.6386200. http://ieeexplore.ieee.org/lpdocs/epic03/wrapper.htm?arnumber=6386200
7. D.M. Ho, J.-S. Hu, J.-J. Wang, Behavior control of the mobile robot for accompanying in front of a human. In *Proceedings of the 2012 IEEE/ASME International Conference on Advanced Intelligent Mechatronics (AIM)* (2012), pp. 377–382. doi:10.1109/AIM.2012.6265891. http://ieeexplore.ieee.org/lpdocs/epic03/wrapper.htm?arnumber=6265891
8. K. Morioka, J.H. Lee, H. Hashimoto, Human-following mobile robot in a distributed intelligent sensor network. IEEE Trans. Ind. Electron. **51**(1), 229–237 (2004). http://ieeexplore.ieee.org/xpls/abs_all.jsp?arnumber=1265801
9. R.C. Luo, N.-W. Chang, S.-C. Lin, S.-C. Wu, Human tracking and following using sensor fusion approach for mobile assistive companion robot. In *Proceedings of the 2009 35th Annual Conference of IEEE Industrial Electronics* (2009), pp. 2235–2240. ISBN 978-1-4244-4648-3. doi:10.1109/IECON.2009.5415185. http://ieeexplore.ieee.org/lpdocs/epic03/wrapper.htm?arnumber=5415185

10. A. Frizera, A. Elias, A.J. Del-Ama, R. Ceres, T.F. Bastos, Characterization of spatio-temporal parameters of human gait assisted by a robotic walker. In *Proceedings of the 4th IEEE RAS & EMBS International Conference on Biomedical Robotics and Biomechatronics* (2012), pp. 1087–1091. ISBN 978-1-4577-1200-5. doi:10.1109/BioRob.2012.6290264. http://ieeexplore.ieee.org/lpdocs/epic03/wrapper.htm?arnumber=6290264http://ieeexplore.ieee.org/xpls/abs_all.jsp?arnumber=6290264
11. C. De La Cruz, R. Carelli, Dynamic Modeling and Centralized Formation Control of Mobile Robots. In *IECON 2006 - 32nd Annual Conference on IEEE Industrial Electronics* (2006), pp. 3880–3885. ISBN 1-4244-0390-1. doi:10.1109/IECON.2006.347299. http://ieeexplore.ieee.org/lpdocs/epic03/wrapper.htm?arnumber=4153091
12. F.N. Martins, W.C. Celeste, R. Carelli, M.Sarcinelli-Filho, T.F. Bastos-Filho, An adaptive dynamic controller for autonomous mobile robot trajectory tracking. Control Eng. Pract. **16**(11), 1354–1363 (2008). ISSN 09670661. doi:10.1016/j.conengprac.2008.03.004
13. F.G. Rossomando, C. Soria, R. Carelli, Autonomous mobile robots navigation using RBF neural compensator. Control Eng. Pract. **19** (3), 215–222 (2011). ISSN 09670661. doi:10.1016/j.conengprac.2010.11.011. http://dx.doi.org/10.1016/j.conengprac.2010.11.011
14. M.W. Whittle. *Gait Analysis: An Introduction*, 4th edn. (Butterworth-Heinemann Elsevier, Oxford, 2003). http://trid.trb.org/view.aspx?id=770947
15. J. Perry, J. Burnfield, *Gait Analysis: Normal and Pathological Function*, 1st edn. (SLACK Incorporated, Grove Road, 1992). ISBN 978-1-55642-192-1
16. M.W. Whittle, D. Levine, Three-dimensional relationships between the movements of the pelvis and lumbar spine during normal gait. Hum. Mov. Sci. **18**(5), 681–692 (1999). ISSN 01679457. doi:10.1016/S0167-9457(99)00032-9. http://linkinghub.elsevier.com/retrieve/pii/S0167945799000329
17. Adept MobileRobots. Pioneer 3-AT (2015). http://www.mobilerobots.com/ResearchRobots/P3AT.aspx
18. SICK. Technical description LMS200 / LMS211 / LMS220 / LMS221 / LMS291 Laser Measurement Systems (2013). http://www.sick-automation.ru/images/File/pdf/LMSTechnicalDescription.pdf
19. C.A. Cifuentes, G.G. Gentiletti, M.J. Suarez, L.E. Rodriguez, Development of a Zigbee platform for bioinstrumentation. In *Proceedings of the 2010 Annual International Conference of the IEEE Engineering in Medicine and Biology Society*, (2010), pp. 390–393. ISBN 9781424441242. http://ieeexplore.ieee.org/xpls/abs_all.jsp?arnumber=5627607
20. C. Cifuentes, A. Braidot, L. Rodriguez, M. Frisoli, A. Santiago, A. Frizera, Development of a wearable ZigBee sensor system for upper limb rehabilitation robotics. In *Proceedings of the 4th IEEE RAS & EMBS International Conference on Biomedical Robotics and Biomechatronics* (2012), pp. 1989–1994. http://ieeexplore.ieee.org/xpls/abs_all.jsp?arnumber=6290926
21. T. Pallejà, M. Teixidó, M. Tresanchez, J. Palacín, Measuring gait using a ground laser range sensor. Sensors. (Basel, Switzerland), **9** (11), 9133–9146 (2009). ISSN 1424-8220. doi:10.3390/s91109133. http://www.pubmedcentral.nih.gov/articlerender.fcgi?artid=3260635&tool=pmcentrez&rendertype=abstract
22. N. Bellotto, H. Hu, Multisensor-based human detection and tracking for mobile service robots. IEEE Trans. Syst. Man Cybern. **39**(1), 167–181 (2009). http://ieeexplore.ieee.org/xpls/abs_all.jsp?arnumber=4695975
23. J. Borenstein, L. Feng, Measurement and correction of systematic odometry errors in mobile robots. IEEE Trans. Robot. Autom. **12**(6) (1996). http://ieeexplore.ieee.org/xpls/abs_all.jsp?arnumber=544770

Chapter 4
Cognitive HRI for Human Mobility Assistance

As aforementioned, in previous approaches of robotics walkers, the user directly commands the robot motion during walking through a HMI. In this context, this chapter presents the implementation and validation of the concept of Cognitive HRI for human mobility assistance. In this approach, the user does not guide directly the walker during walking. In contrast, the walker follows close enough the user in order to provide partial body-weight support. This concept intends to achieve natural human-walker cooperation during the assisted-gait.

This chapter also addresses the integration of the control strategy proposed in Chap. 3 on a robotic walker. That way, some remarks regarding the human-robot physical link demand a new human-walker parameters detection. Consequently, new validations were performed before performing the control implementation. This study is based on a previous work [1].

This chapter is organized as follows. First, the control strategy is presented in the context of human-walker interaction, and a new robotic walker platform is presented to fulfill the sensor and interaction requirements. Second, some experimental trials show the need of developing a new parameter detection algorithm, and a new strategy is formulated and evaluated. Finally, an experimental study is performed to validate both the control parameters detection and the control implementation.

4.1 Interaction Strategy Applied in Smart Walker

The human model interaction presented in Fig. 3.1 is here implemented in a smart walker as it can be seen in the Fig. 4.1. It is noteworthy that variables with regards to the robot now are regarding to the walker. The variables and parameters used in the presented model are: human linear velocity (v_h), human angular velocity (ω_h), human orientation (ψ_h), walker linear velocity (v_w), walker angular velocity (ω_w) and walker orientation (ψ_w). The interaction parameters were defined as the angle φ between v_h and $\overline{WH}$ (named Human-Walker Line), the angle θ between $\overline{WH}$ and

C.A. Cifuentes and A. Frizera, *Human-Robot Interaction Strategies for Walker-Assisted Locomotion*, Springer Tracts in Advanced Robotics 115,
DOI 10.1007/978-3-319-34063-0_4

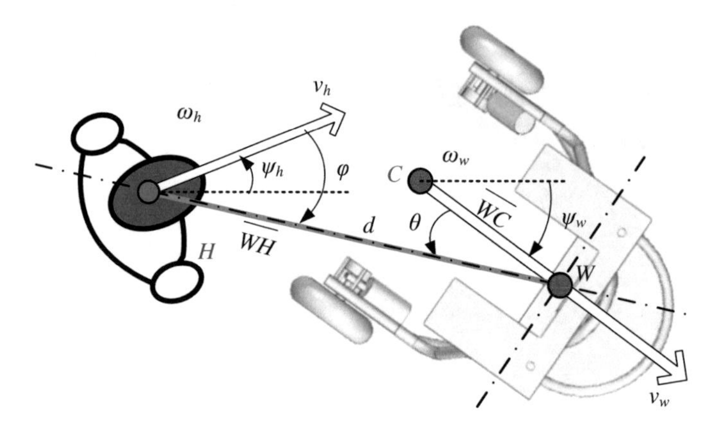

Fig. 4.1 Proposed model for the Human-Walker interaction

$\overline{WC}$ segments, and d, the length of $\overline{WH}$. Finally, the parameter a defines the distance between the controller reference point (W) and the walker center of rotation (C).

The controller Eqs. (3.2) and (3.3) to perform the interaction strategy were previously presented in Chap. 3. In this chapter, the same strategy is implemented in a robotic walker. The next section will describe the robot and sensor setup system that corresponds to a Multimodal-Interaction Platform for human mobility assistance.

4.2 Multimodal-Interaction Platform

This section discusses the hardware and software components of the robotic platform named as *UFES's Smart Walker* (Fig. 4.2). The developed robotic platform consists of a pair of differential rear wheels driven by DC motors and a front caster wheel.

An embedded computer based on the PC/104-Plus standard performs control and processing tasks. It consists of a 1.67 GHz Atom N450 with 2 GB of flash memory (hard disk) and 2 GB of RAM. The application is integrated into a real-time architecture based on Matlab Real-Time xPC Target Toolbox. A laptop computer is used for programming the real-time system and to save the data from the experiments. It is connected to the PC/104-Plus by UDP protocol. If data recording for offline analysis is not necessary, the robotic system is able to operate without the mentioned laptop computer.

The device is designed to provide assistance during the gait based on a multimodal-interaction platform for the acquisition and the real-time interpretation of human gait parameters.

This platform has implemented the modalities previously used in a carrier robot application (see Fig. 3.6). In this approach, legs detection position from the walker,

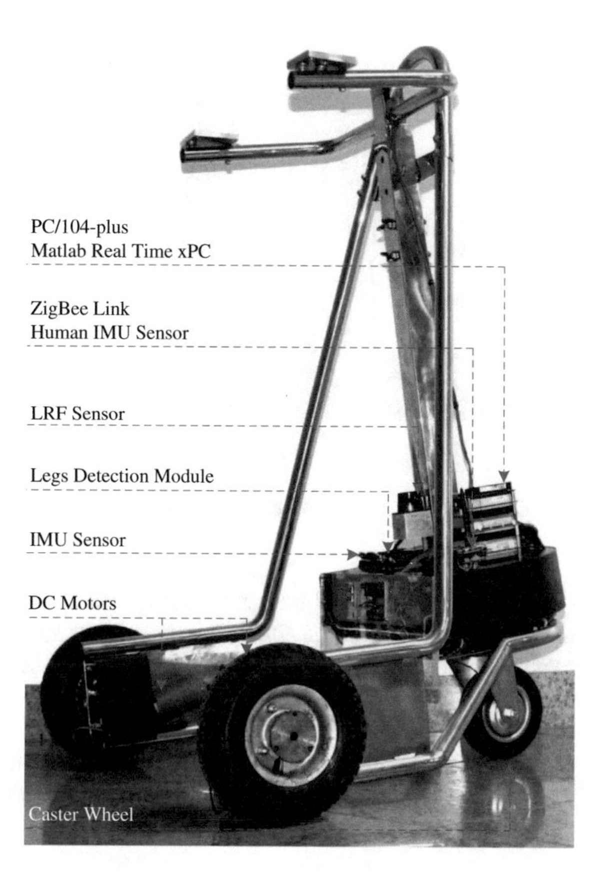

Fig. 4.2 UFES smart walker

human hip and walker orientation are combined to get control parameters described in Fig. 4.1. The sensor configuration is explained as follows.

4.2.1 Leg Detection Module

One LRF sensor Hokuyo URG-04LX [2] is mounted at the legs height, which is used to detect the legs position through the Legs Detection Module (Fig. 4.2) implemented on a processing board based on the dsPIC33F microcontroller. It is linked to the embedded computer by serial interface RS232 and provides the position of each leg every 100 ms.

The leg detection module is used to measure the spatiotemporal parameters of human gait, including θ angle, d, and v_h. The LRF sensor (Hokuyo URG-04LX) is installed on the center column of the walker at a height of 30 cm from the floor. This location allows detecting the user's lower limbs without interference from neither the shoe tip nor the knee.

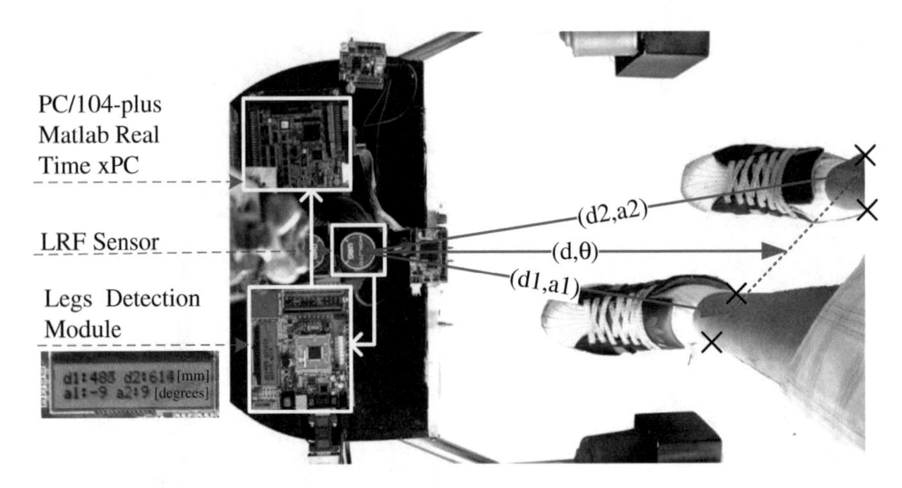

Fig. 4.3 Description of the legs' position detection

The leg detection combines techniques presented in [3, 4] as was presented in the previous chapter (see Fig. 3.6). The legs' positions are calculated in polar coordinates as it can be seen in Fig. 4.3. The general process is based on the differences between two transitions events that define a leg pattern (see x-marks in Fig. 4.3). After that, both distance and angle measurements are calculated in relation to the middle point of each leg. In Fig. 4.3, ($d1$, $a1$) and ($d2$, $a2$) represent the polar coordinates of the left and right legs, respectively. Thus, θ and d control parameters are calculated from the legs' position.

4.2.2 Human Hip and Walker Orientation

Two IMU sensors, developed in previous research [5, 6]. were used. One of these sensors is installed in the walker structure and the other is placed on the user's pelvis (Fig. 4.4). They are linked using ZigBee protocol and communicates via serial interface with the embedded computer. Basically, the IMU information is used to get walker orientation, human orientation (pelvis orientation) and human angular velocity. All data are sampled every 20 ms.

4.2.3 Sensor Readings During Walker-Assisted Gait

As aforementioned, the *UFES's Smart Walker* is used to test the cHRI strategy proposed in this book. Several preliminary tests were performed in order to validate the control parameters detection presented in the previous chapter. These evaluations

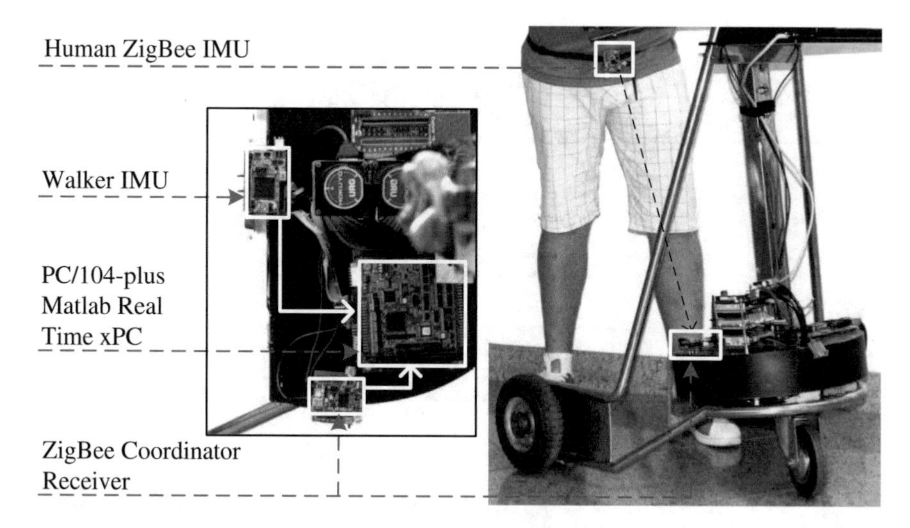

Fig. 4.4 Integration of human IMU and walker IMU sensors

Fig. 4.5 User path used to evaluate the sensor readings during walker-assisted gait

were performed without traction, which means that the user guides the walker as a conventional passive rollator walker. However, the sensor interfaces are enabled to measure the human-walker interaction parameters.

The control parameter detection approach previously defined is based on the zero cross-points detection over the pelvic angular velocity as shown in Fig. 3.8. This temporal information is used to detect when occurs a gait cycle in order to estimate the control parameters, which yield control actions on the robot to follow the user.

As a representative case, Fig. 4.5 shows an experiment when the user performs a specific path, which is composed of three segments, as follows: a left-turn (−90°, Fig. 4.5a), a short straight path (Fig. 4.5b) and a right-turn (90°, Fig. 4.5c).

The sensor readings regarding the user path guiding the walker (Fig. 4.5) are depicted in Fig. 4.6. Figure 4.6a, b show the legs' position during the experiment. These signals correspond to the experiments presented in the previous chapter. Figure 4.6 shows the walker orientation that represents upper-limbs orientation guiding the walker without traction, and the human orientation represents the pelvic orientation. In Fig. 4.6d, both the human and walker angular velocities are shown. It is noteworthy that human angular velocity has not zero cross-points when the user

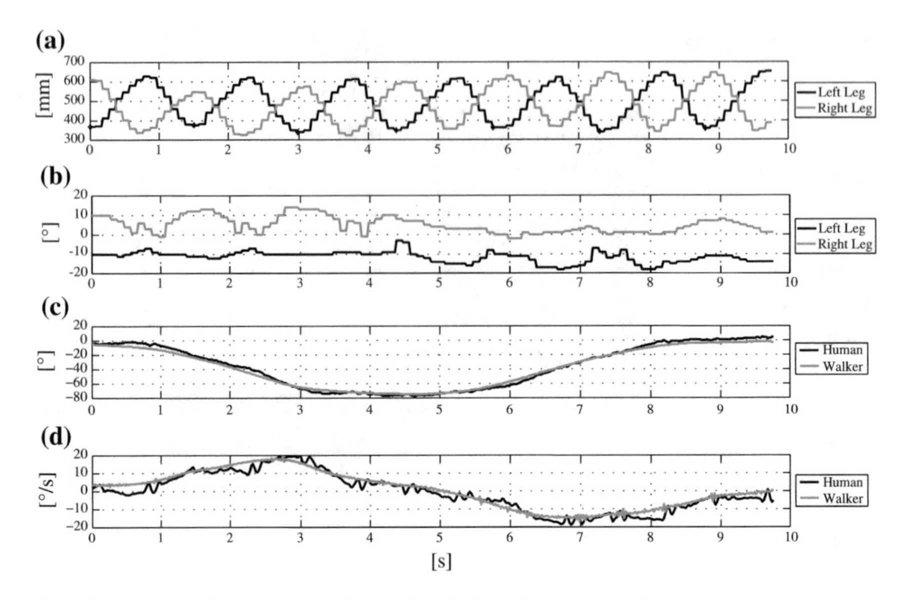

Fig. 4.6 Sensor readings during walker-assisted gait without a control strategy

is making a curve. These zero cross-points are the main source to detect the control parameters in the carrier robot approach (Fig. 3.8, Chap. 3). In contrast, during the walker-assisted gait, the walker angular velocity affects the human angular velocity due to a human-walker physical link. Consequently, it is necessary to propose an alternative methodology for estimating the human-walker control parameters. This strategy will be addressed in the next section.

4.3 Human-Walker Parameters Detection

The new strategy for parameter detection is based on online gait cadence estimation, which is used to continuously estimate the human-walker interaction parameters. The method to obtain the parameters of the proposed model is presented in Fig. 4.7 and described as follows:

1. θ and d are measured directly using the LRF sensor after the legs detection process is performed. Legs Difference Distance (*LDD*) is defined as the difference between the left and right legs distance, which is used for the computation of the v_h. Such detection will be addresses in Sect. 4.3.2.
2. The human linear velocity (v_h) is obtained through the product of gait cadence (GC) and the gait step amplitude (*LDD* amplitude estimation) that are obtained after the leg detection process. Such estimation will be addresses in Sect. 4.3.2.

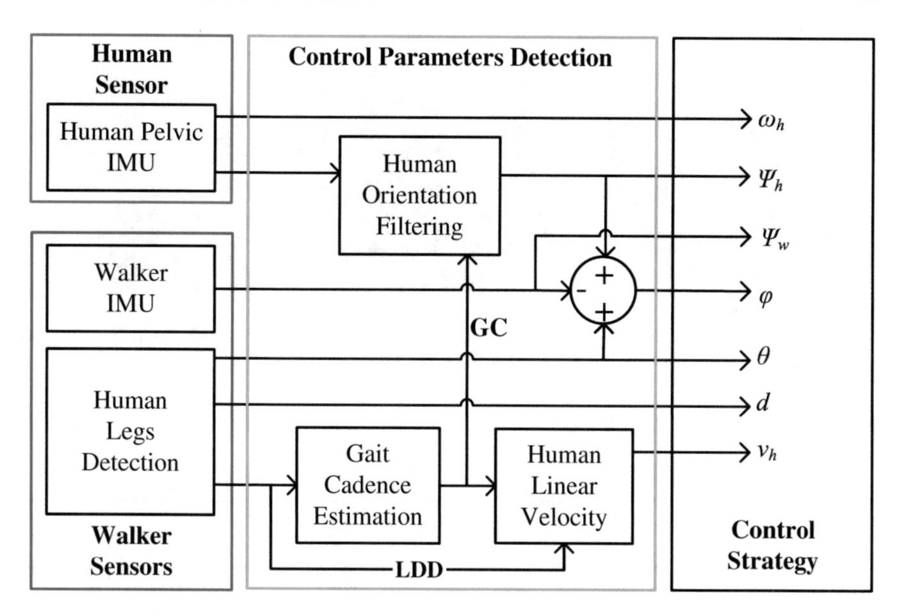

Fig. 4.7 Proposal of a multimodal Human-Walker interaction parameters detection

3. The human angular velocity (ω_h) and orientation (ψ_h) are obtained from one IMU located on human pelvis. The gyroscope integrated on the IMU returns the ω_h measurement from the gyroscope and also returns the yaw angle after the IMU orientation algorithm is performed. ψ_h is also filtered to eliminate the cadence component as a cause of the pelvic rotation during the gait (see Fig. 3.4).
4. The walker orientation (ψ_w) is measured by an onboard IMU, similarly to the previously presented technique.
5. φ represents the orientation difference between v_h and $\overline{WH}$ segment. In Fig. 4.1, φ is equal to $\theta - \psi_w + \psi_h$, and this angle only is defined if the magnitude of the v_h is greater than zero.

4.3.1 Calibration of LRF Sensor

Legs Difference Distance (*LDD*) signal is the reference input of this parameters detection method as it can be seen in Fig. 4.7. *LDD* is defined as the difference between the distances of the left and right legs. It allows the estimation of lower-limbs kinematics parameters and performs the filtering of the oscillatory components contained into the user movement intention, as it will be depicted in next sections.

Due to the fact that the *LDD* signal measured is affected by LRF location, which is placed in a higher plane than the ground, after analyzing the experiments with different users without any dysfunctions associated with gait, a constant ratio K

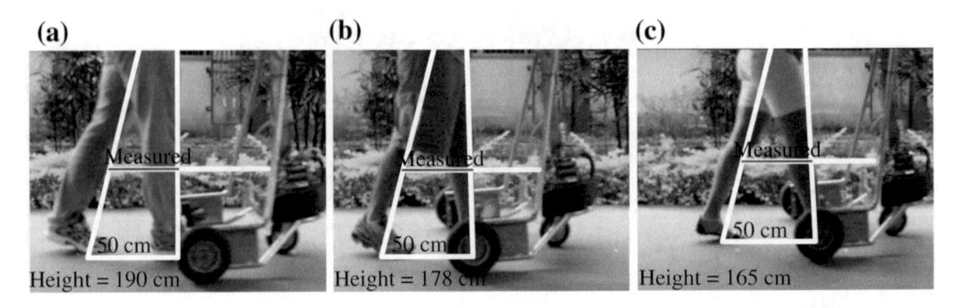

Fig. 4.8 Relationship between actual step length and LRF measurement. **a** Person with *tall height*. **b** Person with *medium height*. **c** Person with *short height*

Table 4.1 Average step length measured during the experiments and average K calculated from each user

User	Height (m)	Cadence (step/s)	Step L. LRF (m)	K
1	1.90	1	0.29	1.71
		0.5	0.33	1.51
2	1.80	1	0.31	1.61
		0.5	0.30	1.65
3	1.78	1	0.31	1.63
		0.5	0.31	1.62
4	1.72	1	0.32	1.58
		0.5	0.31	1.61
5	1.68	1	0.29	1.71
		0.5	0.32	1.54
6	1.65	1	0.30	1.64
		0.5	0.31	1.60

related to LRF measured and step length was obtained (Fig. 4.8). This constant is used to obtain the actual step length.

Six users without any gait dysfunctions were chosen for the estimation of an adequate valor for the K constant. Users heights were between 1.65 and 1.90 m as it can be seen in Table 4.1. Each user performed twice a straight path with the walker; the first one with a half-step-per-second cadence and the second one with a one-step-per-second cadence. A metronome set the pace of the gait, furthermore, steps of 0.5 m were marked with tape on the ground, to help the user to keep a constant step length.

All the experiments were recorded in order to measure the K value for the proposed correction. The average values of the step length measured and the K ratio from each user are presented in Table 4.1. Finally, the average step length of the all users is 0.31 m and the average of the all K values is 1.62. Among the users, 0.29 and 0.33 m were the minimum and maximum values obtained. Adopting a constant K ratio, the error for these users is close to 0.02 m that corresponds to 4 %.

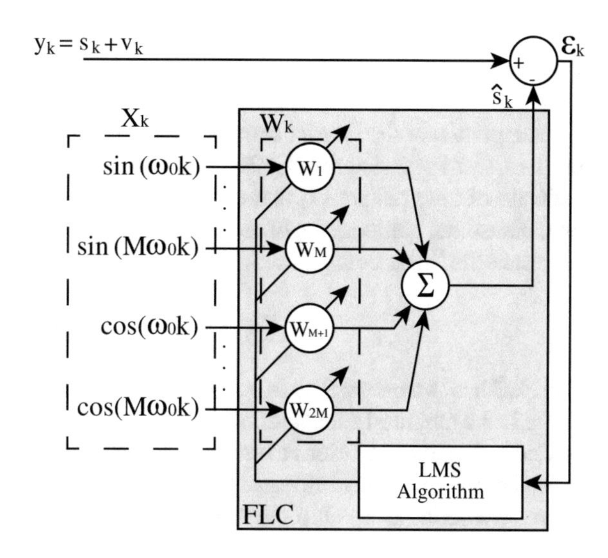

Fig. 4.9 Diagram to illustrate the FLC Algorithm

4.3.2 Adaptive Estimation of Gait Components

The parameters regarding the human gait can be modeled with a Fourier representation due to its periodic nature. This section presents the Fourier Linear Combiner (FLC) and Weighted-Frequency FLC (WFLC) formulations [7]. These tools are here applied to the human gait, in order to estimate gait-related components and perform the filtering of the control parameters, as follows.

FLC estimates both the amplitude and phase of quasi-periodic signals with a known frequency. It operates by adaptively estimating the Fourier coefficients of the model according to the Least Mean Square (LMS) algorithm [7]. The model is based on the M harmonics of the dynamic Fourier model presented in (4.1).

$$s_r = \sum_{r=1}^{M} \left[w_r sin\left(r\omega_0 k\right) + w_{r+M} \cos\left(r\omega_0 k\right) \right] \tag{4.1}$$

FLC algorithm has two inputs as can be seen in Fig. 4.9. The first input is the reference signal (x_k) (4.2) composed of a set of harmonics of the sine and cosine signals with frequency $f_0 = \omega_0/2\pi$.

$$x_{r_k} = \begin{cases} sin\left(r\omega_0 k\right), & 1 \le r \le M \\ \cos\left((r - M)\omega_0 k\right), & M + 1 \le r \le 2M \end{cases} \tag{4.2}$$

The second input of FLC algorithm is (ϵ_k) (4.3), which is the result of the subtraction of the input signal (y_k) and the estimate oscillatory component ($\hat{s_k}$). (y_k) is composed of one oscillatory periodic component that is estimated by the FLC algorithm, plus one stationary input component without oscillatory component (v_k).

$$\varepsilon_k = y_k - \mathbf{W}_k^T \mathbf{X}_k \tag{4.3}$$

The adaptation of the Fourier series coefficients $\mathbf{W}_k$ is performed dynamically based on the Least Mean Square (LMS) recursion, which is a method based on a special estimate of the gradient [7] that ensure inherent zero phase. The harmonic orthogonal sinusoidal components of x_k along with the adaptive weight vector ($\mathbf{W}_k$) (4.4) represent the linear combination.

$$\mathbf{W}_{k+1} = \mathbf{W}_k + 2\mu\varepsilon_k\mathbf{X}_k, \tag{4.4}$$

The FLC has two parameters to be tuned. M is the number of harmonics of the model, and μ is the amplitude adaptation gain. In the approach presented in this work, FLC is used in different situations to estimate parameters based on the gait cadence, which is used as the frequency reference of the presented model. The first application is the adaptive filtering of the hip oscillations to obtain a more stable orientation signal. Furthermore, a FLC-based approach is used for the real-time estimation of human velocity. These FLC applications will be addressed in the next sections.

4.3.3 Gait Cadence Estimation

The GC (Gait Cadence) signal is used as the frequency input signal to obtain the filtering and the estimation of the gait components. This way, this proposal modifies the FLC block (described in the last section) to be useful for GC estimation. It requires a method to adapt the reference frequency to the primary input frequency. This can be done by replacing the fixed frequency (ω_0) of the FLC with an adaptive frequency (ω_{0_k}), which learns the input frequency via an LMS algorithm in the same way that the FLC weights learn the input amplitudes [8]. This approach is known as WFLC (Weighted-frequency Fourier Linear Combiner). Thus, the WFLC recursion minimizes the error ϵ_k between the input s_k and a harmonic model (4.5).

$$\epsilon_k = s_k - \sum_{r=1}^{M} \left[w_{r_k} sin\left(r\omega_{0_k} k\right) + w_{r_k+M} \cos\left(r\omega_{0_k} k\right)\right] \tag{4.5}$$

In this study, it is assumed that the evolution of the difference between the left and right legs distance (*LDD*) can be modeled as a sinusoidal signal of frequency ω_{0_k} plus M harmonics. The WFLC algorithm can be represented as follows:

$$x_{r_k} = \begin{cases} sin\left(r\sum_{t=1}^{k}\omega_{0_t}\right), & 1 \leq r \leq M \\ \cos\left(r\sum_{t=1}^{k}\omega_{0_t}\right), & M+1 \leq r \leq 2M \end{cases} \tag{4.6}$$

In (4.6), x_{r_k} represents a sinusoidal signal with M harmonics with fundamental frequency ω_{0_t}.

The error, which serves to adaptively fit x_{r_k} to the input signal, is described in (4.7).

$$\varepsilon_k = s_k - \mathbf{W}_k^T \mathbf{X}_k - \mu_b \tag{4.7}$$

Frequency and amplitude are updated based on the LMS algorithm expressed in (4.8) and (4.9) [7].

$$\omega_{0_{k+1}} = \omega_{0_k} + 2\mu_0 \varepsilon_k \sum_{r=1}^{M} r \left(w_{r_k} x_{M+r_k} - w_{M+r_k} x_{r_k} \right) \tag{4.8}$$

$$\mathbf{W}_{k+1} = \mathbf{W}_k + 2\mu_1 \varepsilon_k \mathbf{X}_k \tag{4.9}$$

Finally, WFLC has five parameters to be set: M, the number of harmonics of the model, $\omega_{0,0}$, the instantaneous frequency at initialization, μ_0, the frequency update weight, μ_1, the amplitude update weight, and μ_b, and the bias weight that compensates low frequency drifts.

Despite that the WFLC algorithm estimates amplitude, as it adaptively adjusts frequency and amplitude, the correct selection of μ_0 and μ_1 parameters can be a complex task. The WFLC algorithm can be turned to robustly estimate the frequency of the input signal and feeding this information to the FLC algorithm that can robustly estimate the amplitude [8], such as proposed in the parameters estimation that will be addressed in the next section.

In the case when there is not input signal, WFLC algorithm could yield a response with frequency different from zero and amplitude equal to zero [9]. In order to improve these filtering strategies, it is necessary to evaluate the magnitude of the amplitude coefficients $||\mathbf{W}_k||$ to estimate GC components when there is an actual input signal. $||\mathbf{W}_k||$ is considered as an output of the WFLC algorithm as it can be seen in Fig. 4.10.

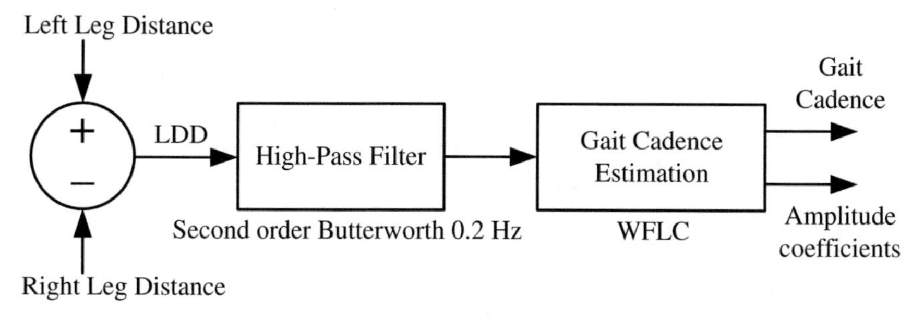

Fig. 4.10 Block diagram to obtain the gait cadence estimation

Furthermore, it is important to perform a previous stage of band-pass filtering (compatible with gait cadence frequencies) allowing a robust adaption to the values of gait cadence and the correct performance of the WFLC [10]. Experimentally, the *LDD* signal has the cadence as the main frequency component, with a High-Pass Filter rejecting static values (DC component) when the user is not walking, due to these events affect the estimation performance. The final algorithm proposed to estimate the GC is shown in Fig. 4.10. The adjustment of the five parameters of this algorithm was obtained experimentally from healthy subjects experimentation ($M = 1$, $\omega_{0,0} = 1$, $\mu_0 = 2 \times 10^{-6}$, $\mu_1 = 1.5 \times 10^{-3}$, $\mu_b = 0$).

4.3.4 Control Parameters Estimation

As previously mentioned, there are two applications that combine FLC and WFLC methods in order to estimate control parameters in this approach. First, in Fig. 4.11, an online adaptive scheme to estimate and cancel the cadence component on ψ_h is shown, which is based on a WFLC block to detect the GC (see Fig. 4.10). The FLC block estimates and subtracts the cadence component on the ψ_h signal. In Fig. 4.11, a gait detector block can also be observed, which gets amplitude coefficients from the WFLC algorithm in order to only perform the filtering when the human is walking. However, the FLC algorithm is always running to avoid transient and adaptation times. To tune the FLC algorithms, the value of the parameters was obtained experimentally for healthy subjects, and $\mu = 0.002$ was the amplitude adaptation gain obtained to filter the orientation. Hip orientation only presents one harmonic ($M = 1$).

The typical orientation values, estimated for ψ_h and ψ_w when the user is guiding the walker in a defined path, respectively from Human (gray) and Walker (black segmented) IMU, are presented in Fig. 4.12. These signals were obtained experimentally without applying any control strategy to the walker's motors. There are three walking segments, and two turn left 90°. As it is shown, the pelvic rotation due to trunk oscillations is contained into the human orientation. The black line (see Fig. 4.12)

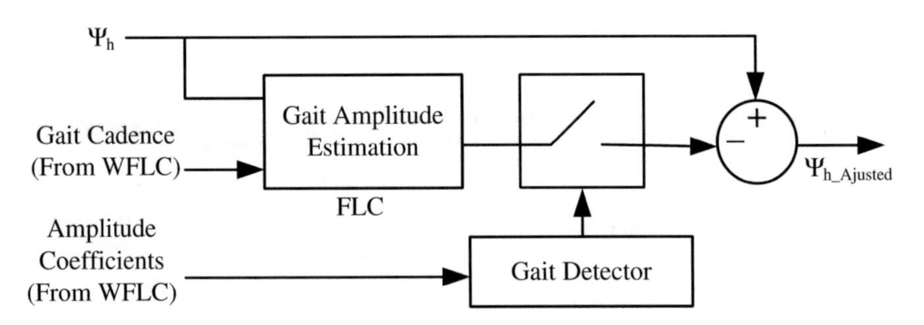

Fig. 4.11 Filtering architecture to cancel the cadence component

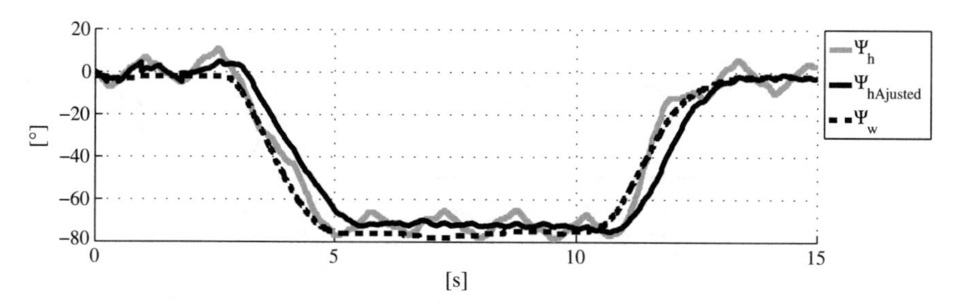

Fig. 4.12 Orientation signals obtained from IMU sensors and human filtered orientation during walker-assisted gait

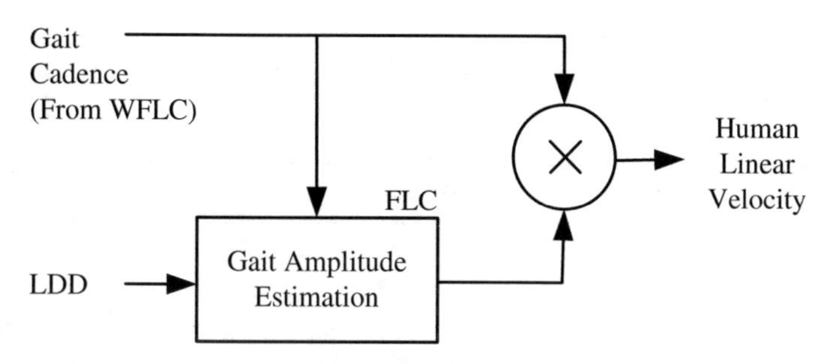

Fig. 4.13 Block diagram to estimate the human velocity

represents the human orientation adjusted after the process of cadence filtering (see Fig. 4.11) is shown.

Human linear velocity can be defined as the product between the GC and the step length [11] and, based on this, the amplitude of the *LDD* signal is estimated as it can be seen in Fig. 4.13. The adjustment of the two parameters of the FLC algorithm was obtained experimentally from healthy subjects ($M = 1, \mu = 0.0018$) to estimated the step length from the *LDD* signal.

As a representative case, Fig. 4.14 shows results of an experiment done with two different velocities. In this case, the human's speed changes during the movement. Speed is changed 500 back to 250 mm/s.

Figure 4.14a shows the distances read by the laser scanner (right and left leg in black and gray, respectively). In Fig. 4.14b, the *LDD* signal is shown, which represents the input for WFLC and FLC algorithms. Figure 4.14c presents the user's GC. Finally, Fig. 4.14d shows the gait amplitude or step length, and Fig. 4.14e shows the obtained linear velocity.

The estimation algorithm shows a tendency to keep constant the distance between the walker and the legs to be used by the control strategy. As a matter of fact, the *LDD* signal oscillates around 0 mm. This indicates that *LDD* is suitable as an input of the WFLC and FLC algorithms.

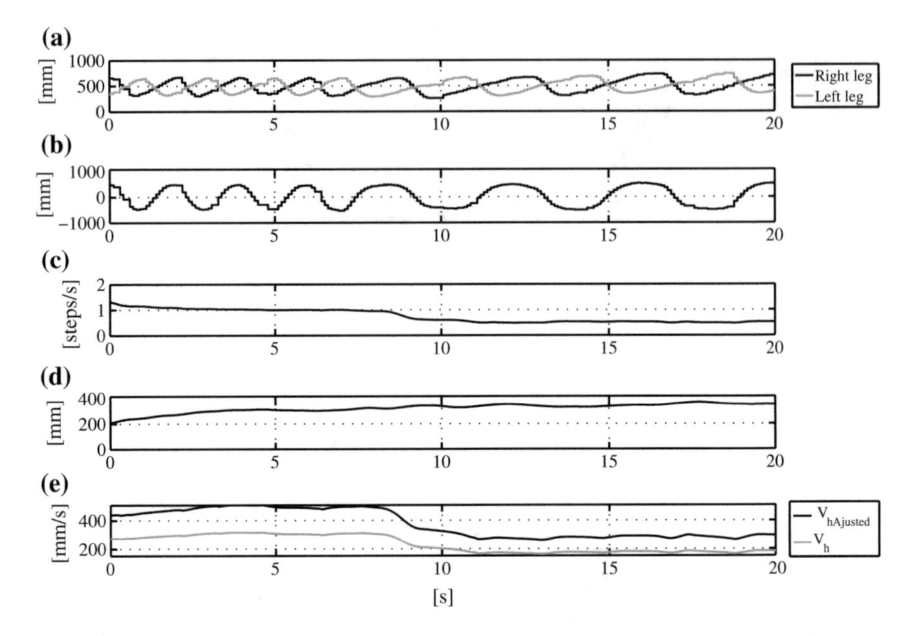

Fig. 4.14 Experiment with speed variation from 500 to 250 mm/s. **a** Legs' position detection from the LRF. **b** LDD. **c** GC estimation. **d** Step length estimation. **e** Human linear velocity

Cadence estimation has adaptive behavior when the user reduces the walking speed, i.e., around one step per second in the first section, which is followed by a section with one step every 2 s. Amplitude is effectively kept close to 300 mm, as expected by the LRF tracking height. Finally, the human linear velocity is around 500 mm/s initially and 250 mm/s in the last part, always with the use of the correction term (K) previously presented in Sect. 4.3.1.

The transitions are observable in all graphs, with the amplitude kept constant. The period of the sinusoidal function describing the legs' distance doubles or halves when switching speed. Such as designed in this HMI strategy, GC and human linear velocity change and reach stable values in approximately 3 s.

4.4 Experimental Study

After processing the LRF sensor calibration, it was observed that the *LDD* signal does not present considerable changes among different healthy (typical) subjects. This signal is the main source to estimate and filter the proposed interaction parameters. Due to this fact, the experimental session was focused on the evaluation of the response of the sensor fusion algorithm, when a user presents changes in both cadence (GC) and step length (SL). This study refers to lower-limbs kinematics parameters for the set of GC, SL and v_h.

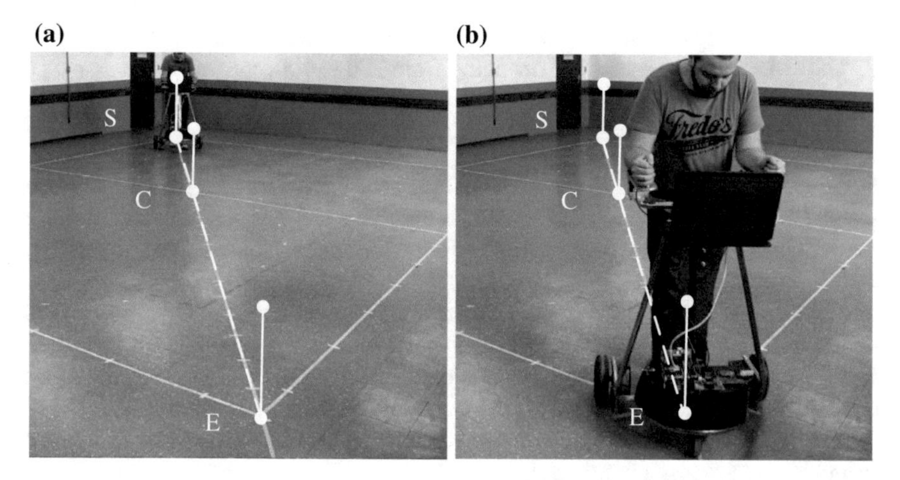

Fig. 4.15 User path guiding the walker (*dashed line*) to evaluate the parameters estimation. **a** Start position performing a straight path. **b** End position performing a straight path

Two different experiments were developed to evaluate the human-interaction parameters without applying any control strategy to the walker's motors. The user can freely drive the walker. Finally, a third experiment was performed with the proposed controller.

In the first experiment, the subject was asked to walk guiding the walker on a straight line marked on the floor. The user was instructed to perform this path with different cadence and step length. A metronome set the pace of the gait. Furthermore, steps of 300 and 600 mm were marked with tape on the ground, to help the user to keep a constant step length. The start (S) and end (E) positions are shown in Fig. 4.15a and b, respectively. In these figures, the center point is also marked (C) as a reference point.

The first experiment contains two parts. In the first part, the user was asked to walk guiding the walker between the point S and E three times with some instructions as follows:

1. Step Length (SL) = 300 mm and Gait Cadence (GC) = 0.6 Steps/s, then velocity = 180 mm/s.
2. SL = 300 mm and GC = 1 Steps/s, then velocity = 300 mm/s.
3. SL = 600 mm and GC = 0.6 Steps/s, then velocity = 360 mm/s.

The selection of these velocities is based on previous experience in human-walker interaction scenarios [12]. Moreover, the second part evaluates the estimation error when changes in the gait kinematics are done. Then, the user is instructed to perform a straight path changing one parameter: SL or GC during each experiment, which is done in the middle of the path. Therefore the user's instructions are changed after crossing the C point, as follows:

1. GC = 0.6 Steps/s constant, SL changing from 300 to 600 mm.
2. GC = 1 Steps/s constant, SL changing from 300 to 600 mm.
3. SL = 600 mm constant, GC changing from 0.6 to 1 Step/s.
4. SL = 600 mm constant, GC changing from 1 back to 0.6 Step/s.

In the second experiment, the user walked with the device with no traction and performed the circle path marked with dashed line. This test was performed to evaluate the proposed technique to estimate the following angular parameters: ψ_w, ψ_h, θ, and φ.

Finally, the third experiment path is an S-shaped path performed twice: once with no traction/controller and once with the proposed control strategy. This was performed to evaluate all interaction parameters when the walker is driven by the user's upper-limbs (no control) and when the walker is following the user without physical interaction by means of the proposed controller, using the parameter estimation methodology proposed in this chapter.

4.5 Results and Discussion

The results of the three experiments show the detection of the human-walker interaction parameters. In addition to that, a comparison of these parameters related to the human guiding the walker without applying any control strategy to the walker's motors and the walker following the human was performed.

4.5.1 Experiment Performing a Straight Path

As a representative case of the obtained results in the first experiment, Fig. 4.16 shows results of a test done with SL = 300 mm and GC = 0.6 Steps/s. In Fig. 4.16a, right and left leg distances are shown in black and grey lines, respectively. Such distances are used to obtain the parameter d. Finally, Fig. 4.16b shows the right and left leg angles (black and grey), which are used to obtain the parameter θ. d and θ are kept constant during the test. Such information represents a comfortable user's position to guide the walker. This information is used for the walker's control strategy to adjust the control set-points and provide a desired position using the smart walker.

LDD signal is represented in Fig. 4.16c, which is the reference signal to estimate the lower-limbs kinematics. In order to compare the estimated parameters with actual values, the *LDD* signal is offline processed to obtain the zero-cross (x) and the maximum (+) points, which are shown in Fig. 4.16c. With this information, the actual values for cadence, step length and velocity per semi-cycle are calculated. The comparisons of actual and estimated values for such parameters are shown in Fig. 4.16d–f.

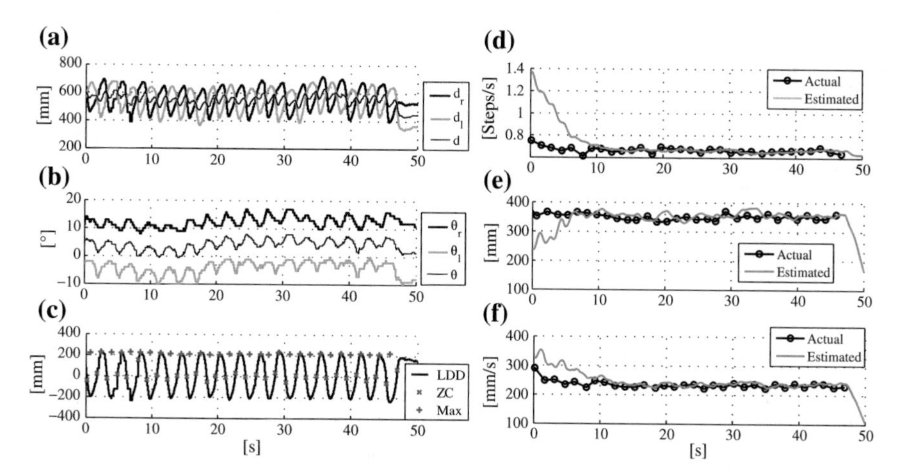

Fig. 4.16 User performing a straight path with SL = 300 mm and GC = 0.6 Steps/s. **a** Legs distance measured and the parameter d obtained. **b** Legs angle measured and θ obtained. **c** *LDD* signal along with maximum and zero-cross points. **d** Actual and estimated cadence. **e** Actual and estimated step length. **f** Actual and estimated human velocity

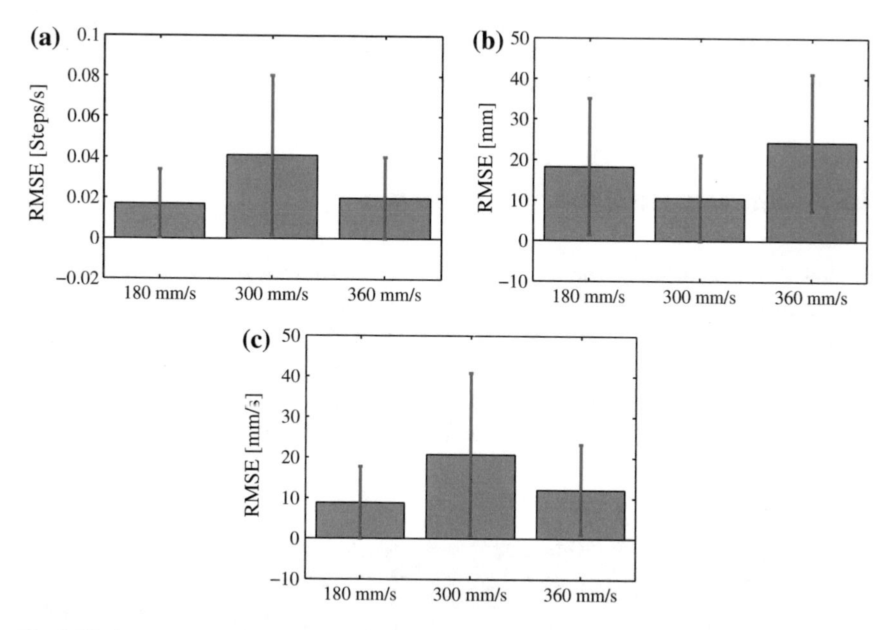

Fig. 4.17 Average errors (RMSE value) of lower-limbs kinematics parameters in experiments with constant step length and cadence. **a** Cadence estimation. **b** Step length. **c** Human velocity estimation

Table 4.2 Average errors (RMSE value) of lower-limbs kinematics parameters in experiments with a change in the parameters performed by the user

	300/600 mm		600 mm	
	0.6 Steps/s	1 Steps/s	0.6/1 Steps/s	1/0.6 Steps/s
Cadence (Steps/s)	0.021	0.020	0.057	0.028
Step L. (mm)	39.8	52.0	16.9	25.3
H. Vel. (mm/s)	15.3 (4.2 %)	16.6 (2.7 %)	28.8 (4.8 %)	26.0 (7.2 %)

An overall analysis of all the data obtained by the first part regarded to the first experiment is shown in Fig. 4.17, which is a chart with root-mean-square errors (RMSE). Typically, the error in cadence remains under 0.05 Steps/s in all tests. Furthermore, the error obtained from each experiment remained close to 5 % as shown in Fig. 4.17a. In Fig. 4.17b, the error in step length estimation remains under 25 mm in all tests. Finally, the human velocity error remains close to 20 mm/s that corresponds to approximately 5 % in each experiment (Fig. 4.17c).

Table 4.2 shows the summary of average errors in experiments with a change in the parameters by the user. The highest error in velocity was lower than 10 % (comparing with the final cadence goal in each experiment), which is adequate to be used in control applications in human-robot interaction [12].

4.5.2 Angular Parameter Evaluation

In the second experiment, the user executed a circle-shaped path while guiding the walker, turning left and right, as shown in Fig. 4.18a and b, respectively.

As a generic representation of the experiment, Fig. 4.19 shows the parameter evolution of one test. Figure 4.19a, b shows the human orientation ψ_h (gray), the filtered human orientation $\psi_h A$ (black), and the walker orientation ψ_w (black segmented). Figure 4.19c, d represents the human angular velocity raw and filtered with a

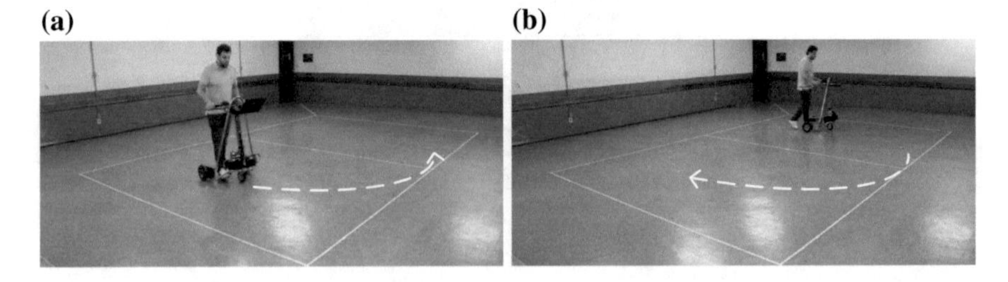

Fig. 4.18 Circle-shaped paths performed in the second experiment. **a** Turning left. **b** Turning right

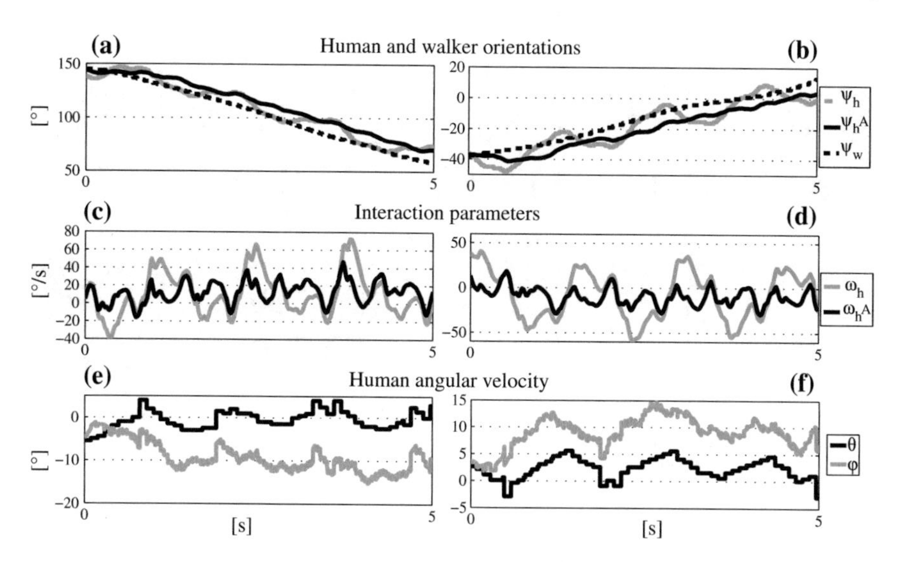

Fig. 4.19 Measurements and estimated parameters in the second experiment. **a** Human and walker orientations while turning *left*. **b** Human and walker orientations while turning *right*. **c** Human angular velocity while turning *left*. **d** Human angular velocity while turning *right*. **e** Interaction parameters while turning *left*. **f** Interaction parameters while turning right

low-pass filter, respectively, in gray and black. Figure 4.19e, f show the θ and φ angles, respectively, in black and gray.

In Fig. 4.19a, b, the result obtained in adaptive filtering of the cadence in the ψ_h signal can be observed. The angle variation of the human follows the variation of the walker because of the intrinsic nature of the human movement, in which turning intention is expressed first on the upper-body and, then, passed to the lower body; the walker orientation is due to the direction imposed by the upper-limbs guiding.

In Fig. 4.19c, d, the filtered human angular velocity (black signal) is positive when turning to the left and negative when turning to the right, which can be used as control parameter. This raw signal (gray signal) contains harmonics not only of the cadence gait but also of the turn intention; hence, a low-pass filter is applied in order to get components related to the turn intention.

In Fig. 4.19e, f, it can be observed that the values of θ (see Fig. 4.1), i.e., the human angular position relative to the walker, are kept between $\pm 15°$, while the user performed turns to the left and to the right. This means that the user does not exit the safe area between the wheels. Moreover, φ, which is one of the control parameters, is negative or positive, when respectively turning left or right, and does not exceed $\pm 5°$ in these experiments.

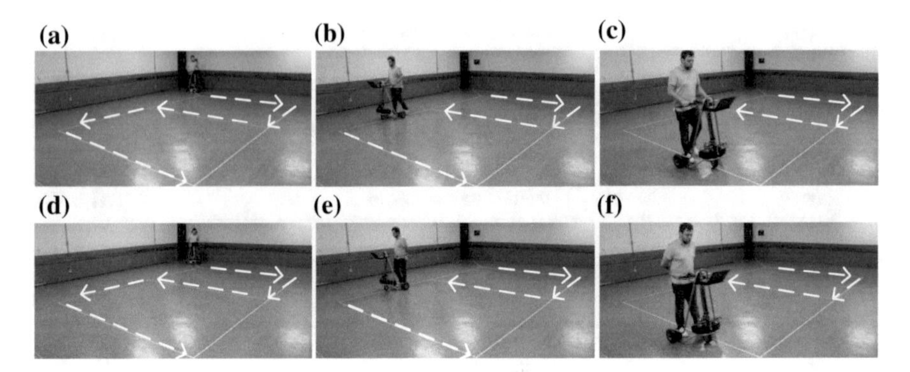

Fig. 4.20 Walker control evaluation performing an s-shaped path. **a–c** Performing an s-shaped path without traction and the user guiding the walker. **d–f** User performing an s-shaped path and the walker following in front

4.5.3 Walker Control System Evaluation

The third experiment is presented in Fig. 4.20a–c. An S-shape path is performed without walker traction, with the human guiding the walker, as in previous experiments. Figure 4.21 represents the evolution of parameters. From the upper to lower sections in Fig. 4.21a, it is possible to distinguish the left leg, right leg, and d in gray, black, and thin black lines, respectively, on the first graph. In Fig. 4.21b, it is possible to distinguish the human linear velocity. Figure 4.21c presents the human and walker orientations ψ_h, ψ_w. Figure 4.21d shows the interaction parameters θ and φ, in black and gray lines, respectively.

The human-robot distance and human linear velocity were kept reasonably close to a constant value of 500 mm and 500 mm/s, respectively. Moreover, it is possible to notice that they evolve very similarly, such as in the previous example. In Fig. 4.21c, the human orientation follows the walker orientation by the fact that the user is guiding the walker with the arms. In this case, the intention of turning is transmitted directly to the device as a motor command. Finally, θ and φ are small values remaining limited between $\pm 5°$ and $\pm 25°$, respectively.

In Fig. 4.20d–f the same path is performed. Nevertheless, the control strategy here proposed is now active. In order to demonstrate the effectiveness of the proposed control, the user has no physical interaction with the device during the test, as shown in Fig. 4.20d–f.

Figure 4.22 shows the evolution of the most significant interaction parameters. It can be observed that the human-robot distance evolves in a very similar way with and without control, meaning that the proposed controller provides natural interaction due to the control law (*d* tends to the desired distance). The desired distance was set at 500 mm, which is based on the experiment with the user guiding the walker. The human velocity was kept reasonably close to a constant value, in this case 400 mm/s.

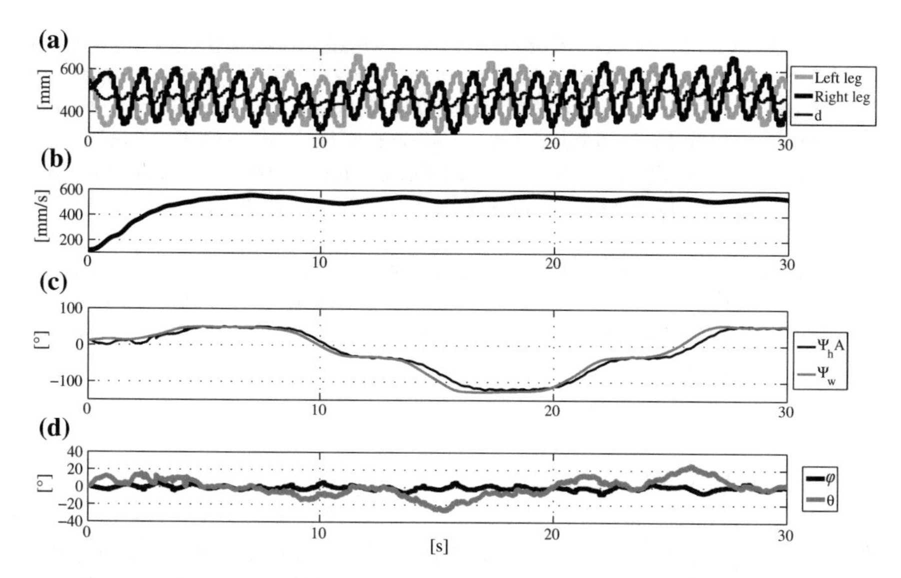

Fig. 4.21 Measurements and estimated parameters performed in an s-shaped path without traction, with the human guiding the walker. **a** Legs' position detection. **b** Human linear velocity estimation. **c** Human and walker orientations. **d** Interaction parameters

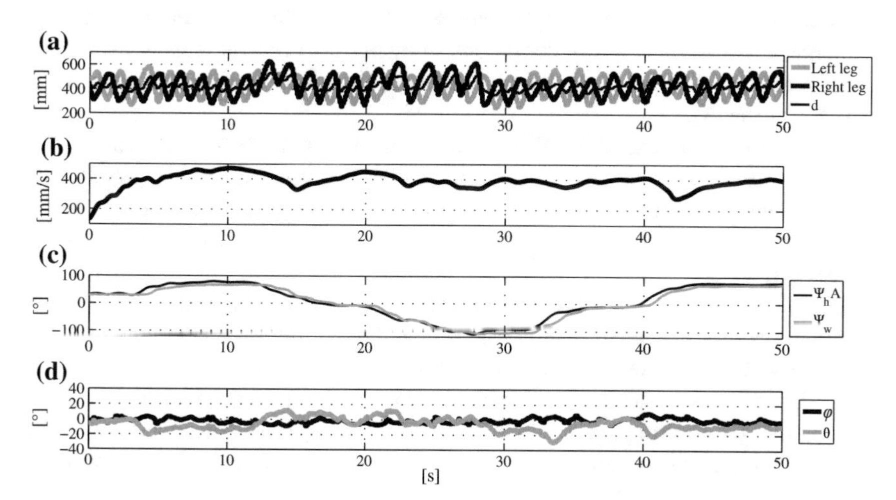

Fig. 4.22 Measurements and estimated parameters performed in an s-shaped path by the user and the walker following in front. **a** Legs' position detection. **b** Human linear velocity estimation. **c** Human and walker orientations. **d** Interaction parameters

The main difference can be observed in the orientation. When the control strategy is used, the walker orientation follows the human orientation. This is due to the fact that the robot is "following in front" of the human, which is yielded by the absence of physical contact. In addition, this orientation sequence between the user and the walker is regarding the human gait performing a turn, which is transmitted in a sequence chain, firstly from the upper-limbs, secondly to the trunk, and finally to the lower-limbs.

Moreover, it can be observed that φ is close $0°$ after every orientation change event due to the control law (φ tends to zero). As in the previous case, θ and φ are limited to small values, such as $\pm 5°$ and $\pm 25°$, respectively, showing that the controller provides natural interaction.

4.6 Chapter Conclusions

This chapter has presented a new multimodal proposal for a human-walker interaction strategy based on the sensor integration of a wearable IMU and an onboard IMU and an LRF. In addition, a new human-walker controller for "following in front" of the user, with an experimental validation of the controller, has been presented.

In the experimental study, despite the continuous oscillation during the walking, the parameter estimation was consistent, showing also repeatability with human linear velocities' changes. In the same way, lower-limbs kinematics estimation errors were lower than 10 %.

This chapter has also showed that the parameter estimation proposed and the control strategy can be effective in guiding a walker to follow a human. The controller keeps the walker continuously following in front of the human during the gait, and it can be observed that the robot orientation follows the human orientation during the real experiments.

One of the advantages of the human-walker interaction proposed in this book is the computational efficiency. The sensor processing algorithms and human tracking from the walker are executed in real time, also showing stable performance. The reliability of this approach is guaranteed with the integration of the analysis of human walking into the control parameters, which includes practical experimentation of the proposed interaction controller, showing the performance of the control system and the parameter detection strategy.

The next chapter will integrate a force interaction subsystem to improve the proposed multimodal interfaces. As observed in the experiments without the use of the proposed controller, it is interesting to integrate upper body force interaction to the strategy, in order to obtain a more predictive behavior.

References

1. C.A. Cifuentes, C. Rodriguez, A. Frizera-Neto, T.F. Bastos-Filho, R. Carelli, Multimodal human - robot interaction for walker-assisted gait. IEEE Syst. J. 1–11 (2014)
2. Hokuyo Automatic Co. Scanning Laser Range Finder URG04LX Specifications. Technical report (2005) http://www.hokuyo-aut.jp/02sensor/07scanner/download/products/urg-04lx/data/URG-04LX_spec_en.pdf
3. T. Pallejà, M. Teixidó, M. Tresanchez, J. Palacín, Measuring gait using a ground laser range sensor. Sensors (Basel, Switzerland) **9**(11), 9133–9146 (2009). ISSN 1424-8220. doi:10.3390/s91109133. http://www.pubmedcentral.nih.gov/articlerender.fcgi?artid=3260635&tool=pmcentrez&rendertype=abstract
4. N. Bellotto, H. Hu, Multisensor-based human detection and tracking for mobile service robots. IEEE Trans. Syst. Man Cybern. **39**(1), 167–181 (2009). http://ieeexplore.ieee.org/xpls/abs_all.jsp?arnumber=4695975
5. C.A. Cifuentes, G.G. Gentiletti, M.J. Suarez, L.E. Rodriguez, Development of a Zigbee platform for bioinstrumentation, in *Proceedings of the 2010 Annual International Conference of the IEEE Engineering in Medicine and Biology Society* (2010), pp. 390–393. ISBN 9781424441242. http://ieeexplore.ieee.org/xpls/abs_all.jsp?arnumber=5627607
6. C. Cifuentes, A. Braidot, L. Rodriguez, M. Frisoli, A. Santiago, A. Frizera, Development of a wearable ZigBee sensor system for upper limb rehabilitation robotics, in *Proceedings of the 4th IEEE RAS & EMBS International Conference on Biomedical Robotics and Biomechatronics* (2012), pp. 1989–1994. http://ieeexplore.ieee.org/xpls/abs_all.jsp?arnumber=6290926
7. B. Widrow, S.D. Stearns, *Adaptive Signal Processing* (Prentice Hall, Englewood, 1985)
8. C.N. Riviere, N.V. Thakor, Modeling and canceling tremor in human-machine interfaces. IEEE Eng. Med. Biol. **15**(3), 29–36 (1996)
9. W.T. Latt, U.X. Tan, K.C. Veluvolu, C.Y. Shee, W.T. Ang, Real-time estimation and prediction of periodic signals from attenuated and phase-shifted sensed signals, in *IEEE/ASME International Conference on Advanced Intelligent Mechatronics, AIM* (2009), pp. 1643–1648. doi:10.1109/AIM.2009.5229825
10. A.F. Neto, J.A. Gallego, E. Rocon, Extraction of user's navigation commands from upper body force interaction in walker assisted gait. BioMed. Eng. OnLine **9**(1), 37 (2010). ISSN 1475-925X. doi:10.1186/1475-925X-9-37. http://www.pubmedcentral.nih.gov/articlerender.fcgi?artid=2924341&tool=pmcentrez&rendertype=abstract, http://www.biomedcentral.com/content/pdf/1475-925X-9-37.pdf
11. M.W. Whittle, *Gait Analysis: An Introduction*, 4th edn. (Butterworth-Heinemann Elsevier, Oxford, 2003). http://trid.trb.org/view.aspx?id=770947
12. A. Frizera, A. Elias, A.J. Del-Ama, R. Ceres, T. Freire Bastos. Characterization of spatio-temporal parameters of human gait assisted by a robotic walker, in *Proceedings of the 4th IEEE RAS and EMBS International Conference on Biomedical Robotics and Biomechatronics* (IEEE, 2012), pp. 1087–1091. ISBN 978-1-4577-1200-5. doi:10.1109/BioRob.2012.6290264. http://ieeexplore.ieee.org/lpdocs/epic03/wrapper.htm?arnumber=6290264, http://ieeexplore.ieee.org/xpls/abs_all.jsp?arnumber=6290264

Chapter 5
Multimodal Interface for Human Mobility Assistance

The previous two chapters presented the implementation of the cHRI strategy for mobility assistance in the context of both mobile robots and smart walkers. In order to complete the physical and cognitive HRI for walker-assisted gait defined in Fig. 2.3, this chapter introduces the physical HRI block. Afterwards, the cHRi and pHRi are integrated into a Multimodal Interface for Human Mobility Assistance. Additionally, some concepts regarding control strategies based on interaction forces during the walker-assisted gait are described and implemented.

This chapter is organized as follows. First, upper-limb reaction forces are implemented as a pHRi. That way, 3D forces sensors are integrated into the UFES's Smart Walker (presented in the previous chapter) by means of forearm supporting platforms. Such modality yields important information for motion control of robotic walkers. However, the review presented in Sect. 2.2.2.3 showed some issues related to the extraction of the upper-limbs guiding intentions, which are addressed in this chapter.

Secondly, a Multimodal Interface for Human Mobility Assistance is presented. This interface integrates LRF, IMU and 3D forces sensors. Consequently, new human walker interaction parameters are presented in order to monitor several body parts during the walker-assisted gait. This interface can be useful as a tool for understanding the human motion intentions. In addition, such parameters can be used as control set-points to evaluate several control strategies for specific gait disorders in order to improve the body weight support during the walker-assisted gait. This information also enables natural channels of communication between the walker and the human.

Finally, some strategies for forces interaction control and a final control strategy are presented. Such control strategy combines concepts of pHRI and cHRI, which are based on both forearm reaction forces and gait kinematics from the legs scanning localization.

C.A. Cifuentes and A. Frizera, *Human-Robot Interaction Strategies for Walker-Assisted Locomotion*, Springer Tracts in Advanced Robotics 115,
DOI 10.1007/978-3-319-34063-0_5

5.1 Integration of an Upper-Limb Interaction Forces System in the Walker Platform

During normal gait, the HAT (Head, Arms and Trunk) is considered to travel as a unit; it moves with the body's center of gravity and also transmits the walking direction to the lower limbs [1]. As a consequence of this, a direct sensing of the HAT orientation is suitable to get the user's walking direction to guide the robotic walker. The HAT orientation can be measured from the interaction between upper-limbs and walker, which produces forces related to the user's partial body weight support.

This section presents the integration of the sensor subsystem for measuring the upper-limb reaction forces in the UFES's Smart Walker. Therefore, a new multimodal sensor configuration to acquire and estimate the human-walker interaction parameters is presented as shown in Fig. 5.1.

Summarizing, this interface is based on a set of sensors: (1) a LRF (Laser Range Finder) sensor is used to detect the legs' distance/position in relation to the walker, (2) a wearable IMU (Inertial Measurement Unit) was adopted to capture the hip orientation, (3) Two 3D force sensors measure the upper-limb interaction forces between the human and the walker. This information represents the guiding forces supplied from the passenger unit (Fig. 5.4c) [2]. The sensor acquisition is performed by an embedded computer onboard the walker. It is based on the PC/104-plus standard and performs the sensor processing according to the sensor fusion strategy.

The system designed to measure the upper-limb interaction forces is shown on Fig. 5.2. The forearm supporting platforms were designed to provide a support area from the elbow up to the hands that is more comfortable and stable than using handlebars. They also stabilize the trunk and upper-limbs during the walker-assisted gait, providing better body-weight support and improving the interaction with the device. The measurement system is composed of two 3D force sensors MTA400 (Futek, US) and six amplifier modules CSG110 (Futek, US). These sensors are integrated under each forearm-supporting platform, which allows measuring six independent components of interaction forces during the assistive-gait. According to the axis configuration (Fig. 5.2), x-axis, y-axis and z-axis represent lateral direction, forward direction and vertical component (including the supported user's body weight), respectively.

In order to understand the force signals during walker assistive-gait, a user was asked to walk freely with the walker. The user was instructed to perform a specific path, such as: (1) performing a straight path, and (2) performing a curve (90°). The signals obtained from the 3D force sensor are depicted in Fig. 5.3.

Despite of the noise included into force signals, it is possible to infer four walking gestures. In (1), at the beginning of the experiment, all forces signals do not present any activity (equal to zero). When the user supports his forearms on the walker's platform, z-axis force becomes more negative, indicating that the user's body weight is partially supported by the walker. In (2), the user is walking in a straight path, y-axis force shows the user intention to go forward, z-axis force depicts that the user is unloading a partial body weight and also the trunk motion component is added. In (3), the user is performing a curve, and y-axis signal shows a positive

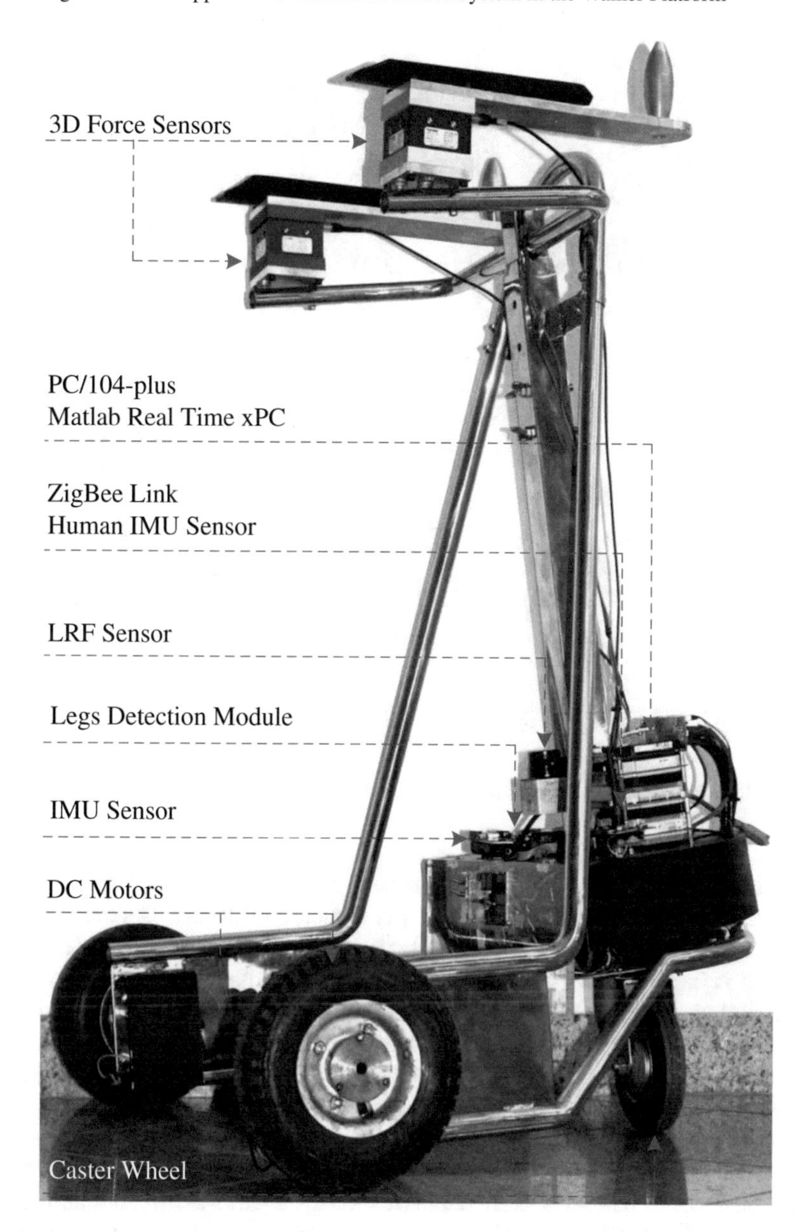

Fig. 5.1 Sensor modalities developed to characterize the walker-assisted gait

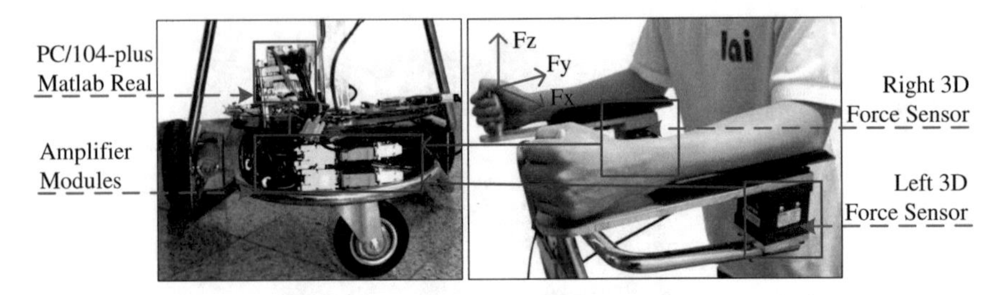

Fig. 5.2 Upper-limb interaction forces acquisition system

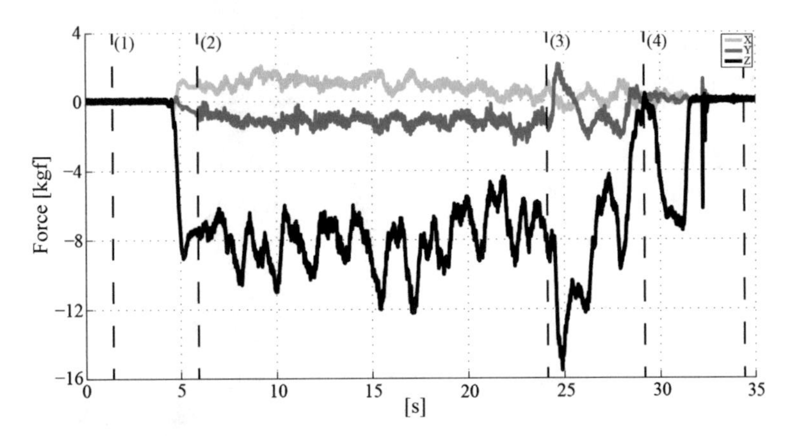

Fig. 5.3 Raw forces signals obtained from the right 3D force sensor during walker-assisted gait

pick that represents a turn intention (Left turn, see Fig. 5.3). At the same time, the body weight support is incremented as it is depicted in the z-axis signal. Finally, in (4) the user stops to walk and the signals return to zero, as the user leaves the device. It is important to mention that x-axis is not considered representative as far as locomotion commands are concerned. Previous works show that this laterality component is strongly correlated to the lateral displacements of the body's Centre of Gravity [3]. Therefore, this work does not consider the x-axis as a source of human-walker interaction parameters for guidance purposes.

In Fig. 5.3, it is possible to observe the presence of higher frequency vibrations into the force components. Such components are caused by the ground-wheel interaction. Additionally, other periodical components related to the trunk oscillations during gait are also found in the force signals. Therefore, a filtering strategy is implemented to obtain the force components that represent the user's guiding intentions, which will be addressed in the next section.

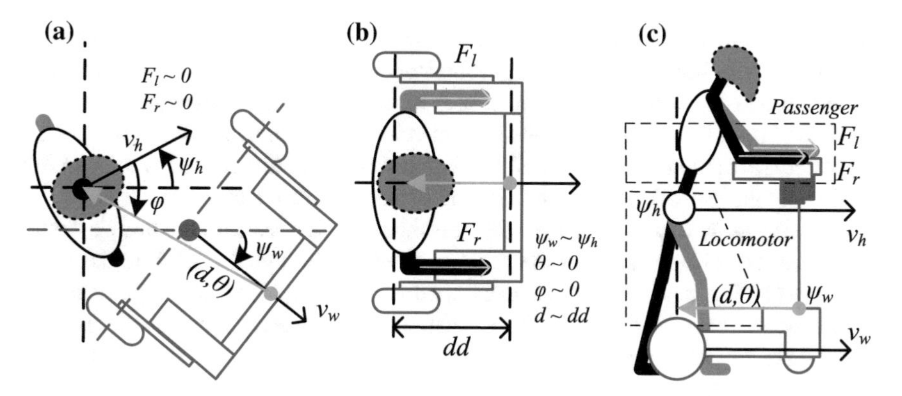

Fig. 5.4 Model of human-walker interaction. **a** Model of human-walker interaction. **b** Human-walker desired position during the walking (dd) and representation of interaction forces (F_l and F_r). **c** Passenger and locomotor units

5.2 Multimodal Interaction Strategy

Due to the fact that human locomotion is not only characterized by legs movement, but instead, it involves coordinated movements in several body parts, it would be desirable to monitor other segments during human motion. This would result in a more predictive and natural human-walker interaction, aiming at a better multimodal interface. Consequently, this section presents a new interaction strategy based on the concepts of cHRI and pHRI presented in this book.

Figure 5.4a shows the parameters that can be useful for cHRI, as presented in the previous chapter (Fig. 4.1): human position from walker point of view (d, θ and φ), human velocity (v_h) and human orientation (ψ_h). In that case, the user does not have physical contact with the forearms support platforms, so the upper-limb forces (F_l, F_r) signals are close to 0 Kgf. However, the cHRI strategy will aim to achieve a comfortable position to promote the body weight support during the walking.

Moreover, when the user has a comfortable position using the walker (see Fig. 5.4b), the cHRI strategy is operating to keep this walker's position from the user. Consequently, the control errors defined in the cHRI strategy return values close to zero. However, the upper-limbs guiding forces return measurements related to both body weight support and the guiding intentions forces. Such information can be used as a control input to enhance the human-walker interaction.

These parameters were classified into three categories to enable different control features and to develop a more natural human-walker interaction.

1. *Human relative position to the walker*: this information is useful to keep the human within a desired position and orientation from the walker's point of view (d, θ, φ), which aims at providing a comfortable user's position to improve the human-walker interaction during the walking. Such parameters may be also applied to obtain better human weight support.

2. *Lower-limbs kinematics*: Among the several parameters, the step length, gait cadence and human velocity (v_h) (obtained from the product between the step length and the gait cadence) are useful to develop control strategies where the walker velocity may adapt to the human velocity. Kinematic changes on the gait patterns are common with age. However, most of the studies found in the literature do not take into account the gait kinematics as a control parameter.
3. *Human movement intention*: the passenger unit moves with the body's center of gravity and also transmits the walking direction through the pelvis to the lower limbs [1] (Fig. 5.4c). As a consequence, a direct sensing of the passenger (from the interaction between the upper-limbs and the walker: F_l and F_r) and hip orientation (ψ_h) are suitable to get the user's walking direction to guide the robotic walker.

In summary, the integration of both *human relative position to the walker* parameters and *lower-limbs kinematics* parameters as inputs of the walker's control strategy could achieve the human-walker desired position during the walking (Fig. 5.4b). Within this approach, d is equal to dd (desired distance), θ is equal to $0°$, the walker velocity (v_w) is equal to v_h and walker orientation (ψ_w) is equal to ψ_h, so φ is close to $0°$. Thus, the physical interaction is generated and *human movement intention* parameters can be used in the walker control strategy to promote an adequate support to performing turns.

5.2.1 Multimodal Interface for the Estimation of Human-Walker Interaction Parameters

The selection of the specific architecture for the multimodal interface for online estimation of human interaction parameters was based on a detailed analysis of the information that can be extracted from each sensor, paying a special attention to the benefits and drawbacks for each choice. The implementation of the interface was as illustrated in Fig. 5.5, which presents the method used to integrate the sensor modalities here proposed (inputs) and the parameters estimation (outputs).

As presented in the previous chapter, the human gait may be modeled as a Fourier representation due to its periodic nature. According to this, authors decided to use FLC (Fourier Linear Combiner) and WFLC (Weighted-Frequency Fourier Linear Combiner) formulations to design and implement the sensor fusion strategy [4]. The detailed formulations of FLC and WFLC were also presented in the previous chapter. In this chapter, the filtering strategy used for obtaining the human orientation was adapted to extract the components related to upper-limbs guiding intentions. The implementation of this sensory architecture allows the estimation of the following parameters that could be used for the control of the smart walker:

1. *Human relative position to the walker*: From the direct measurement of the legs detection system, d and θ parameters are defined. Additionally, the φ parameter is obtained combining both the leg detection system and the IMU sensors. They

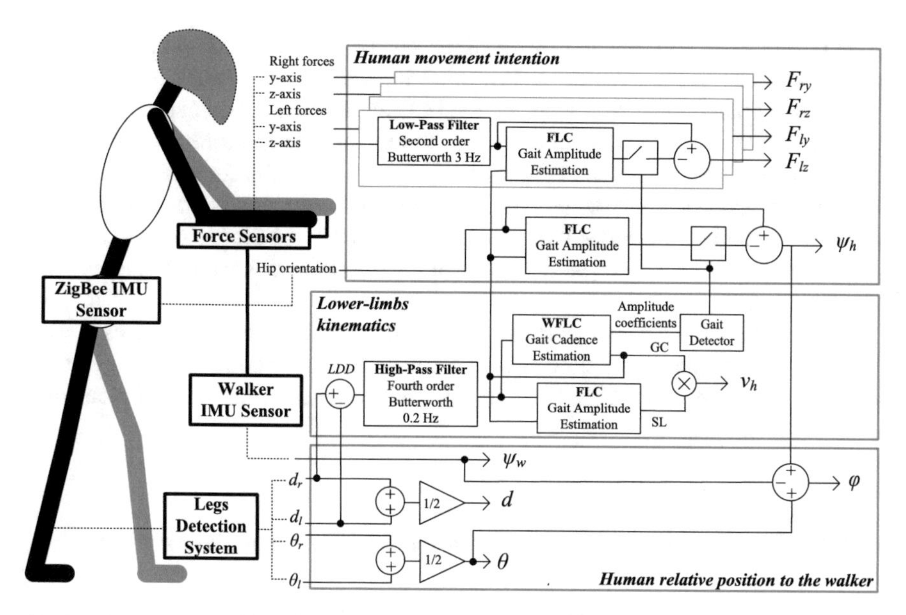

Fig. 5.5 Diagram that illustrates the multimodal interface for online estimation of human interaction parameters in walker-assisted gait

represent the human relative position from the walker. This information is useful in control strategies to keep the human with a desired position using the walker.

2. *Lower-limbs kinematics*: Human velocity is obtained by the product between the Gait Cadence (GC) with the step-length amplitude estimation as shown in Fig. 5.5. The FLC algorithm was implemented in order to get a robust estimation of the step length amplitude (Sect. 4.3.2). The WFLC algorithm used to the estimation of GC was introduced in Sect. 4.3.3.
3. *Human movement intention*: it is estimated based on hip orientation and upper-limb guiding forces. The filtering strategy for the estimation of these variables is based on an independent on-line adaptive schemes to estimate and cancel the cadence component on the hip orientation and the upper-limb guiding forces, as it can be seen in Fig. 5.5. These signals are:
   (a) Hip orientation, obtained from the yaw angle returned from the IMU located on the human pelvis.
   (b) Right and left arm y-axis forces, representing the forward directions from arms.
   (c) Right and left arm z-axis forces, user's body weight supported on the force sensors.

The FLC block is designed to estimate and cancel cadence component of each input signal, which receives the gait cadence (frequency input) obtained from the

WFLC block. In Fig. 5.5, it can be observed a gait detector block, which gets amplitude coefficients from the WFLC algorithm in order to perform the filtering when the human is walking. However, the FLC algorithm is always running to minimize the transients and adaptation delays. To tune the FLC algorithms, the value of the parameters was obtained experimentally for healthy subjects, and ($\mu = 0.002$) was the amplitude adaptation gain obtained to filter the forces and orientation. However, forces signals presented two harmonics ($M = 2$) and the hip orientation only one harmonic ($M = 1$). The performance of the filtering architecture is addressed in the next section.

As commented in related works [2, 3, 5], the upper-limb reaction forces present frequency components introduced by ground-wheel interaction. This way, at the beginning of the force filtering schemes, a low-pass filter was implemented (see Fig. 5.5).

5.2.2 *Evaluation of Human Movement Intention Parameters*

The parameters' validation of both *human relative position to the walker* and *lower-limbs kinematics* was presented in Sect. 4.4. Consequently, this section presents an experimental validation regarding the user movement intention parameters. Two different experiments were developed in order to verify the accuracy in the estimation of such parameters. A volunteer without any dysfunctions associated with gait was chosen to perform the experiments. This user performed changes in the cadence and the step length to estimate adaptive filtering of the hip orientation and the upper-limbs guiding intention. Moreover, continuous turns in the user path were performed in order to simulate real human locomotion scenarios. The performance of such algorithms would not change among different subjects, as observed in [5].

5.2.2.1 Experimental Study

In the first experiment, the subject was asked to walk guiding the walker on a straight line marked on the floor. The user was instructed to perform this trajectory with different cadence and step length. A metronome set the pace of the gait. Furthermore, steps of 300 and 600 mm were marked with tape on the ground in order to help the user to keep a constant step length. The start (S) and end (E) positions are shown in Fig. 5.6a.

The user was asked to walk guiding the walker between the point S and E three times with some instructions as follows:

1. Step Length (SL) = 300 mm and Gait Cadence (GC) = 0.6 Steps/s (velocity = 180 mm/s).
2. SL = 300 mm and GC = 1 Steps/s (velocity = 300 mm/s).
3. SL = 600 mm and GC = 0.6 Steps/s (velocity = 360 mm/s).

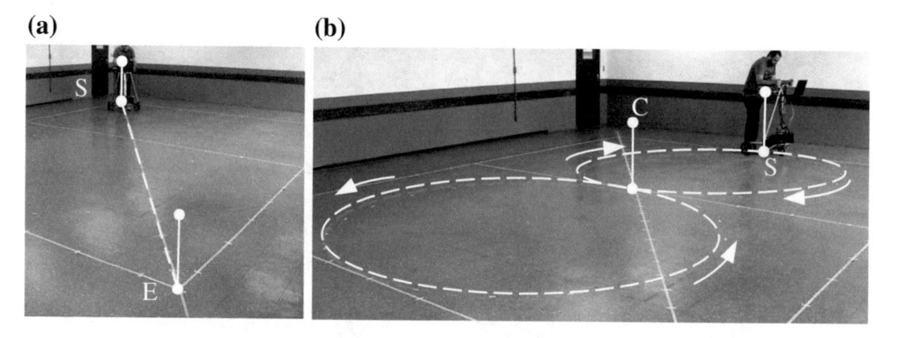

Fig. 5.6 User path guiding the walker (*dashed line*) to evaluate the parameters estimation. **a** Performing a straight path. **b** Performing an *eight-shaped* curve (lemniscate)

In the second experiment, the user is asked to perform an eight-shaped path (lemniscate) as shown in Fig. 5.6b. This experiment is performed to evaluate and understand parameters such as: the guiding intentions from the upper-limb reaction forces and the human hip orientation.

5.2.2.2 Results and Discussion

As a representative case of the obtained results in the first experiment, Fig. 5.7 shows the parameters evaluated during a test (SL = 300 mm and GC = 0.6 Steps/s) performing a straight-line path. Such parameters are the hip orientation and the upper-limb guiding forces. Raw forces (R), force signals after low-pass filtering (LP) and filtered signals (F) are illustrated in Fig. 5.7a–d, which correspond to right y-axis, left y-axis, right z-axis and left z-axis forces, respectively. Frequency analysis was performed offline using FFT algorithm. Despite the low forces applied on the y-axis (Fig. 5.7a, b), when a straight-line path is executed, the rejection of the cadence components on the force signals is performed. Particularly, in this experiment two harmonics were rejected as shown in Fig. 5.7f, g. However, the filtering effect is more evident on the z-axis as it can be seen in Fig. 5.7h, i, which is a cause of high forces yield by the partial body weight unloading.

Orientation angles are shown in Fig. 5.7e. The hip angle obtained from IMU (R), filtered hip orientation (F) and cadence rejection on this signal are observed in Fig. 5.7j. Finally, the filtered hip orientation and the walker orientation (W) are highly related to each other, as expected.

Comparing the actual and filtered values in the frequency domains, a high accuracy is obtained by the fact that the low frequency components (guiding intentions) were kept despite the high rejection of the cadence components. A filter quality indicator (FQI) was defined in order to measure the cadence rejection components (5.1). This indicator compares both spectrum data, raw and filtered, and was evaluated into interval of the cadence values detected during each experiment.

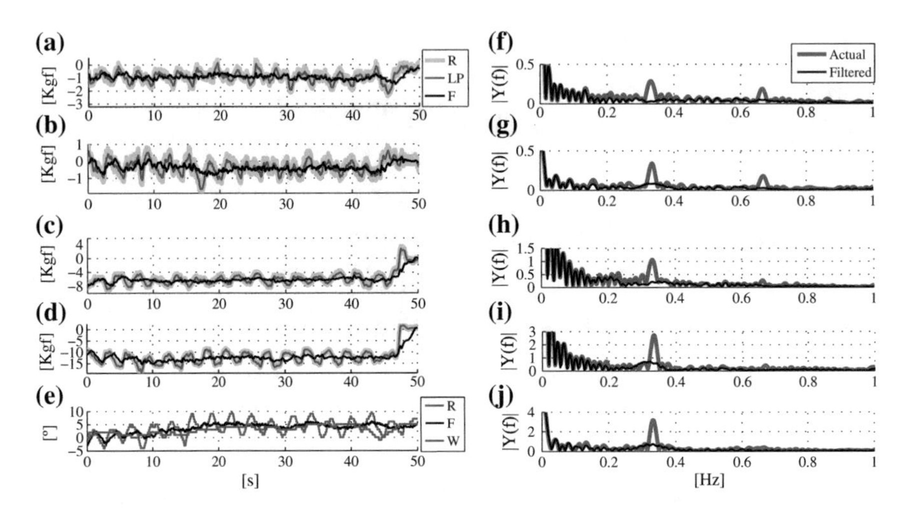

Fig. 5.7 Temporal data and frequency spectrum of hip orientation and upper-limb guiding forces performing a straight path (raw signal (R), signal after low pass filter (LP), filtered signal (F) and walker signal (W)). **a**, **f** Right arm y-axis force. **b**, **g** Left arm y-axis force. **c**, **h** Right arm z-axis force. **d**, **i** Left arm z-axis force. **e**, **j** Orientation angles

Table 5.1 Filter quality indicator of hip orientation and upper-limb guiding forces in experiments with constant step length and cadence performed by the user

FQI	300 mm 0.6 Steps/s	300 mm 1 Steps/s	600 mm 0.6 Steps/s
Right z-axis	0.730	0.643	0.651
Left z-axis	0.710	0.752	0.664
Right y-axis	0.745	0.745	0.687
Left y-axis	0.727	0.688	0.669
Hip orientation	0.746	0.789	0.667

$$FQI = 1 - \frac{\sum_{k=f_{1,1}}^{f_{1,2}} \left|Y_{Filtered(k)}\right| + \sum_{k=f_{2,1}}^{f_{2,2}} \left|Y_{Filtered(k)}\right|}{\sum_{k=f_{1,1}}^{f_{1,2}} \left|Y_{Actual(k)}\right| + \sum_{k=f_{2,1}}^{f_{2,2}} \left|Y_{Actual(k)}\right|} \tag{5.1}$$

$f_{1,1}$ and $f_{1,2}$ are the minimum and maximum values that were estimated for the first harmonic of gait cadence. $f_{2,1}$ and $f_{2,2}$ define the second interval according to the second estimated harmonic. Table 5.1 shows the results after applied the FQI into the signals of hip orientation and upper-limb guiding forces. These signals were obtained from the experiments when the user was asked to walk with constant cadence and step length as follows. There were not significant differences among hip orientation and guiding forces signals. In the same way, the experimental conditions do not affect the adaptive filtering scheme. The mean rate of cadence components rejection was 71 % over the experiments presented in Table 5.1.

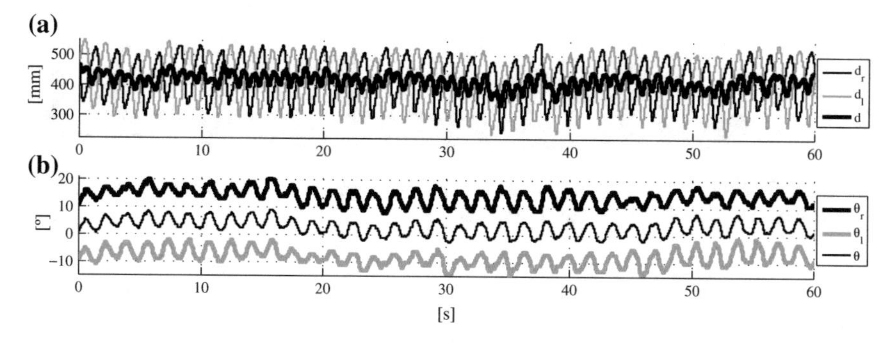

Fig. 5.8 Human relative position to the walker performing an *eight-shaped* curve. **a** Measured Legs' distance and d parameter obtained. **b** Measured Legs' angles and θ parameter obtained

The second experiment was developed to understand the guiding intention components extracted from the upper-limbs reaction forces and the hip orientation. These signals are intended to interface human intention with trajectory control of the smart walker. Figure 5.8 shows the legs' location detection performing the eight-shaped curve proposed in the third experiment (Fig. 5.6b). Despite continuous curves in the human trajectory, the legs positions detection relative to the walker were stable; even these measurements present the same fashion performing a straight path. Therefore, this information does not present any relevant characteristics related to be performing a curved trajectory. However, it confirmed that these signals are useful to estimate the lower-limbs and gait kinematics parameters.

A Spectrogram study was performed to compare the raw and filtered signals as can be seen in Fig. 5.9. As a representative case, left arm reaction force is only depicted in

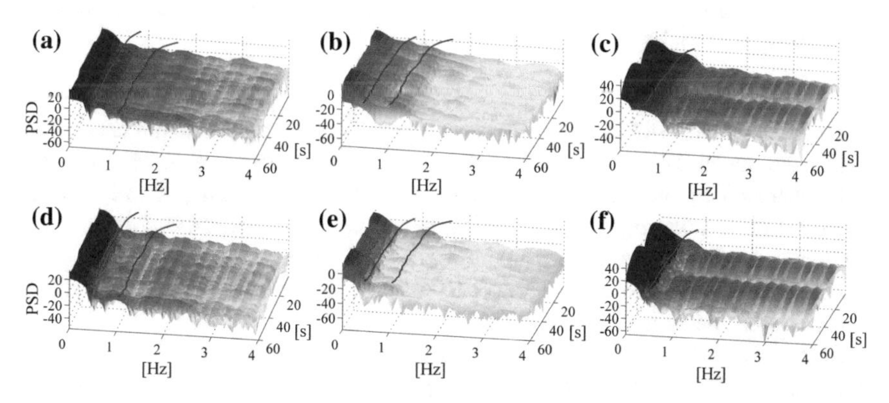

Fig. 5.9 Power spectral density regarding the signals obtained from both the left arm reaction forces and the hip orientation performing an *eight-shaped* curve. The *solid blue lines* represent the harmonics estimated (WFLC) to be canceled by means of the filter adaptive schemes (FLC). **a** Raw z-axis force. **b** Raw y-axis force. **c** Raw hip orientation. **d** Filtered z-axis force. **e** Filtered y-axis force. **f** Filtered hip orientation

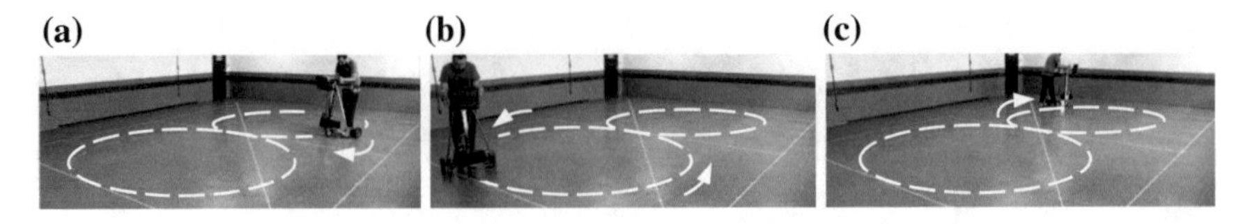

Fig. 5.10 Still images of the second experiment performing an *eight-shaped* curve (lemniscate). **a** First semicircle. **b** Circle path. **c** Last semicircle path

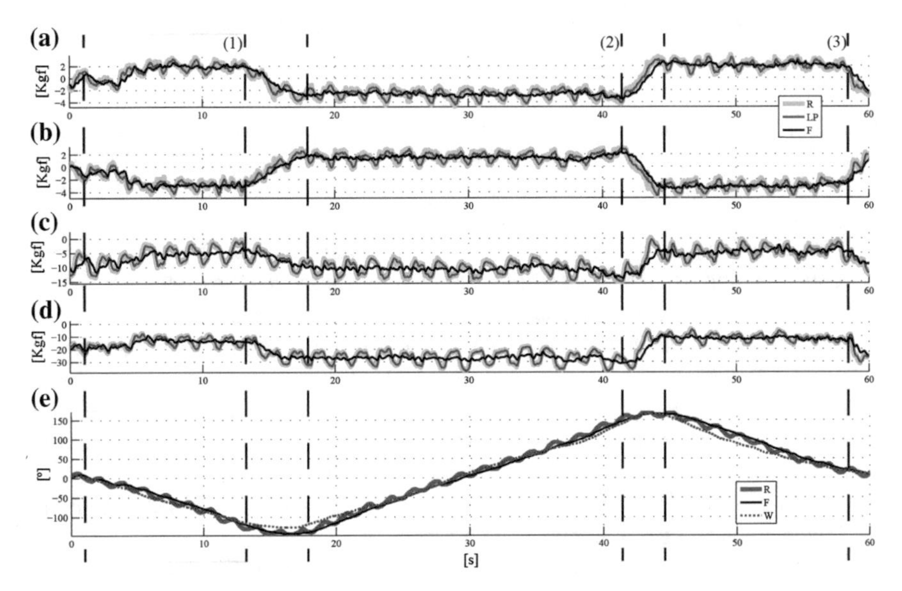

Fig. 5.11 Temporal data of hip orientation and upper-limb guiding forces performing an eight-shaped curve (raw signal (R), signal after low pass filter (LP), filtered signal (F) and walker signal (W)). **a** Right arm y-axis force. **b** Left arm y-axis force. **c** Right arm z-axis force. **d** Left arm z-axis force. **e** Orientation angles

Fig. 5.9a, b, d and e. In Fig. 5.9c, f, the spectrogram results applied to hip orientation signals are shown. The solid blue lines represent the frequency components estimated to be canceled by means the proposed adaptive filtering scheme, see Fig. 5.5. In general, with this filtering strategy, the low frequency components that correspond to the human guiding intentions are not affected. Furthermore, it is possible to observe that the cadence related components are rejected despite the complex trajectory (eight-shaped) performed by the user.

Some still images of the second experiment are shown in Figs. 5.10, and 5.11 shows temporal data of upper-limb guiding forces and hip orientation performing an eight-shaped curve (lemniscate). It is noteworthy that the eight-shaped path is analyzed in three phases that define specific guiding intentions: (1) a first semicircle path (user turning right) Fig. 5.10a; (2) a circle path (user turning left) Fig. 5.10b; and (3) a last semicircle path (user turning right) Fig. 5.10c.

In the first phase, after the sensor fusion strategy exceeds the transient (almost 5 s), the right arm forces on the y-axis and z-axis are close to 2 and −5 Kgf, respectively. As a result, the arm left forces on the y-axis and z-axis are close to −3 and −13 Kgf, respectively. These values depend on both each user and asymmetrical support that can be normally found in the interaction frameworks. However, the forces magnitudes and directions are maintained during right turning gesture, which can be also ensured in the third phase. Moreover, the hip orientation decreased and followed the walker orientation while the user was turning right. It is important to mention that the walker turns before the human hip because the walker is guided by the upper limbs. During left turning, the hip orientation increases and follows the walker orientation. Furthermore, the right arm forces on the y-axis and z-axis are close to −3 and −11 Kgf, respectively. At the same time, the left arm forces on the y-axis and z-axis are close to 2 and −28 Kgf, respectively. These values were maintained while the user was performing the second phase.

Finally, it is important to mention the sequential response of the parameters signals between the first and second phases: first, a falling edge and a rising edge were generated by the right y-axis force and left y-axis force, respectively. Second, the hip orientation slope was equal to zero, and the force signals reached the typical values of the second phase. Finally, the hip orientation slope begins to increase up to the typical value in this phase. This sequence is also possible to detect between the second and third phases.

This information allows analyzing the motion chain when the user is turning from the passenger to locomotor units. The combination of these signals could enhance the control strategies to support the user during the turn. Thus, the forces signals are useful to produce fast commands to guide the wheels during the turn. It was observed by the falling and rising edges that the user performs up to get the target orientation. Moreover, the hip orientation does not present an important activity during the turn (slope close to zero). However, it allows understanding that the user is performing the turn with the upper-limbs. At the same time, the hip is in a quasi-static position and the legs are in double-support stage. Finally, when the user gets the orientation target, the motion is transmitted from the upper-limbs to the lower-limbs by the hip. In this case, the walker provides a static support to maintain the body equilibrium.

5.3 Strategies for Forces Interaction Control in Robotic Walkers

Basically, the forces interaction controllers translate forces and/or torques signals into walker's velocities. A review of the literature concerning such control strategies showed that two approaches were tested with users with mobility dysfunctions [2, 6–8], which are explained as follows.

The first method is the admittance control, which consists of virtual mass and damper parameters to provide natural and intuitive interaction between user and device. The mass-damper model acts as a low pass filter so that the high frequency noise due to shock, gait cadence and vibration from the system can be reduced. The damping parameter returns the output to equilibrium as quickly as possible without oscillating [6].

Indeed, the admittance control approach allows the walker's dynamics to be set like a linear or nonlinear system, subject to limitations of actuator power, servo control bandwidth, and computation limitations. Models with fast dynamics require higher bandwidth and fast sampling time for the control system. Complex models obviously require more computation power, however, these do not appear to be significant issues for devices with slow motions [7].

Finally, the second method is divided in two parts: first, the extraction of forces components related to user's intention motion, which are the input of a fuzzy controller (second part). Due to the fact that extracted components represent clearly a proportional signals regarding the upper-limb guiding intentions, a basic controller based on fuzzy rules can be implemented. Consequently, the motor intention generates velocity set-points as a function of the user's motor intention. Such method does not take into account the walker's dynamics, but it offers smooth and responsive motor commands as is shown in [2, 8].

The multimodal interface proposed in this research returns the guiding intention forces in a proper way as it was validated in the previous section. The next section presents a control strategy that includes a forces interaction controller based on the second method, which generates velocity set-points as a function of the user's motor intention. In fact, the modality that extracts the upper-limb guiding intentions along with a fuzzy controller represents the pHRI block defined in Fig. 2.3.

5.4 Example of a Controller Based on pHRI + cHRI

This section describes a control strategy based on physical and cognitive HRI, which is shown in Fig. 5.12. This strategy is divided in two different controllers. On the one hand, a pHRI controller based on fuzzy logic in which the inputs are the guiding intention components extracted from the upper-limbs reaction forces (F_l and F_r), and the output is the walker's angular velocity (ω_w). It is noteworthy that the forces signals are useful to produce fast commands to guide the wheels during the turn, and the hip orientation does not present an important activity during the turn as was presented in Fig. 5.11. Consequently, the IMU located on the human hip is not used in this approach, so this control strategy does not require any sensor to be attached to the user's body, which can be considered an improvement in the usability.

On the other hand, a cHRI controller based on inverse kinematics was added as can be seen in Fig. 5.12, the control error is $\tilde{d}$, the input controlled is v_h, and the walker's linear velocity v_w is the action control. That control proposal integrates the concept of robot following in front of the user, which was presented in the previous chapter. Indeed, the kinematics controller here implemented is a reduced structure of the controller proposed in Fig. 3.2.

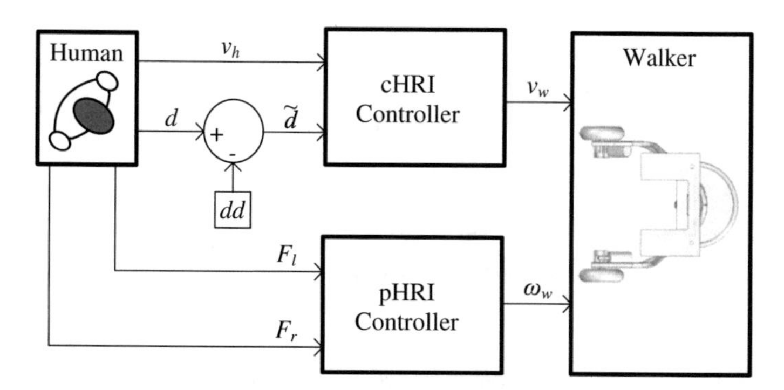

Fig. 5.12 Block diagram of the proposed controller based on cHRI and pHRI

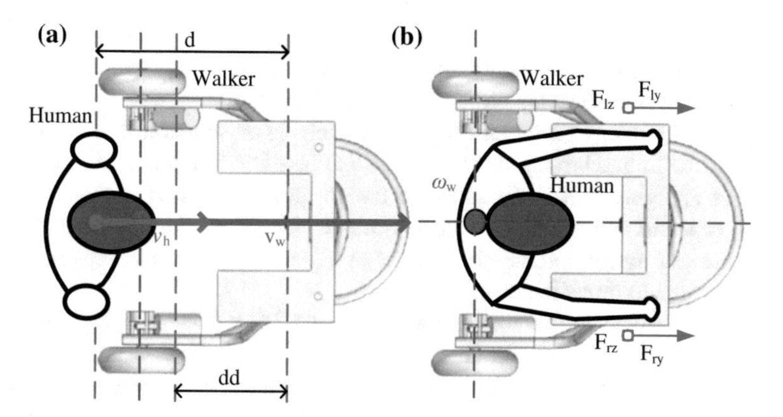

Fig. 5.13 Control implementation based on cHRI and pHRI. **a** Human-walker model to obtain v_w. **b** Human-walker model to obtain ω_w

5.4.1 *Control Implementation*

Figure 5.13a shows the strategy to control walker linear velocity (v_w). The variable to be controlled is the human-walker distance d. The control objective is to achieve a desired human-walker distance $d = dd$. This also reaches to improve the human-walker physical interaction.

The expression $\dot{\tilde{d}} = -v_h + v_w$ shows the basic direct kinematics of the walker, where $\tilde{d} = d - dd$ (Fig. 5.13a) is the difference between the desired and the measured distance. Therefore, the inverse kinematics controller is $v_w = v_h - k\tilde{d}$, where k is a positive gain to be adjusted. Such controller equation corresponds to a specific case of the controller presented previously in (3.2) with $\theta = 0$ and $\varphi = 0$.

As aforementioned, the upper-limbs reaction forces are suitable inputs to guide the orientation of the robotic walker. y and z force components from right and left sensors are filtered individually using the filtering architecture previously presented (Fig. 5.5). From Fig. 5.13b, F_{ly} and F_{ry} present proportional values to the movement

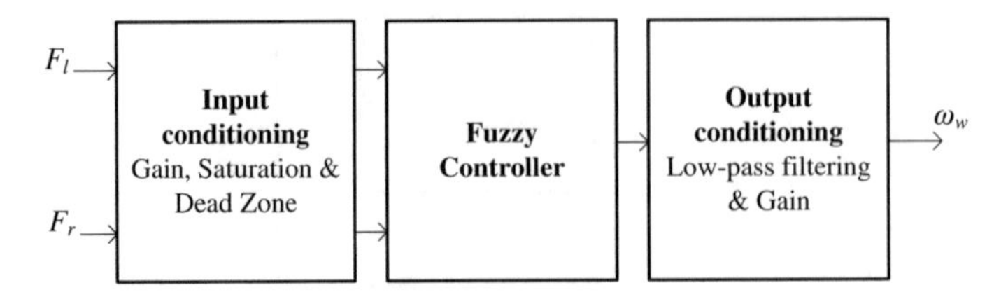

Fig. 5.14 Angular velocity controller base on user interactions forces

intention. These components were divided by the z-components (F_{lz} and F_{rz}) in order to obtain force signals (F_l and F_r) that are also proportional to the amount of the body weight applied on each armrest. This feature is important in cases of asymmetrical support caused by a unilateral affection on gait patterns. These filtered forces are used in this approach to drive the walker angular velocity (ω_w) through a classifier and controller based on fuzzy logic (Fig. 5.14).

The control strategy based on force interactions is presented in Fig. 5.14. Then, signals are conditioned to input the fuzzy logic classifier. The conditioning process consists of applying a *gain*, to adjust to the correct range of inputs; a *saturation function*, to avoid values over the input limits of the fuzzy classifier; and a *dead-zone*, to prevent motor commands in cases of signals very close to zero and, thus, not high enough to move the device.

The main element of the control scheme (Fig. 5.14) is the fuzzy logic block. It is built upon the information obtained experimentally from the tests performed with healthy subjects. It combines information of right and left sensors to generate angular velocity commands. The filtered and conditioned force signal inputs can vary from -1 to $+1$ and are grouped into five classes as can be seen in Fig. 5.15a:

- $Negative_{High}(N_H)$, *Z-shaped* function with $a = -0.8$ and $b = -0.3148$, Eq. (5.2).

$$zmf(x) = \begin{cases} 1, x \leq a \\ 1 - 2 \cdot (\frac{x-a}{b-a})^2, a \leq x \leq \frac{a+b}{2} \\ 2 \cdot (b - \frac{x}{b-a})^2, \frac{a+b}{2} \leq x \leq b \\ 0, x \geq b \end{cases} \tag{5.2}$$

- $Negative_{Low}(N_L)$, Gaussian symmetrical function with $\sigma = -0.1173$ and $c = -0.4$, Eq. (5.3).

$$gaussmf(x) = e^{\frac{-(x-c)^2}{2\sigma^2}} \tag{5.3}$$

- $Zero(Z)$, Gaussian symmetrical function with $\sigma = 0.2045$ and $c = 0$.
- $Positive_{Low}(P_L)$, Gaussian symmetrical function with $\sigma = 0.1173$ and $c = 0.4$.
- $Positive_{High}(P_H)$, *S-shaped* function with $a = 0.8$ and $b = 0.3148$, Eq. (5.4).

$$smf(x) = \frac{1}{1 + e^{-a(x-b)}} \tag{5.4}$$

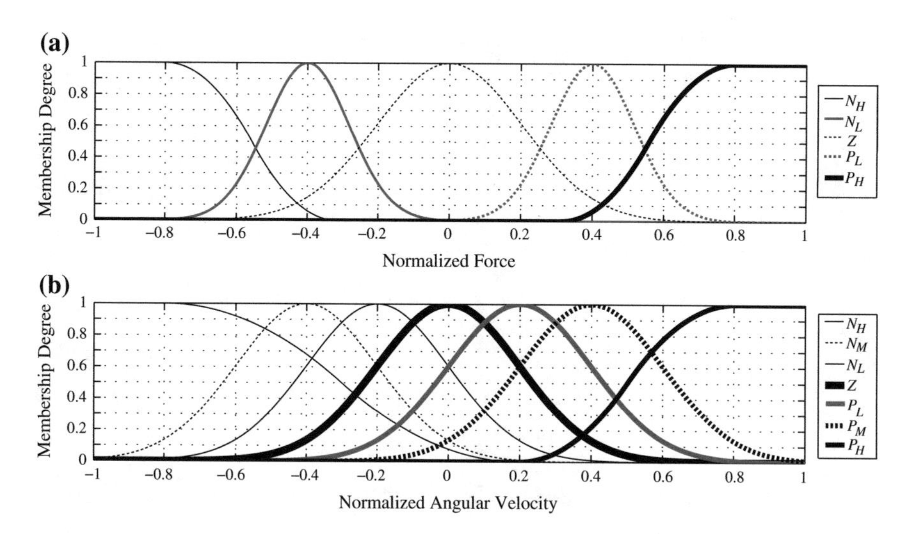

Fig. 5.15 Membership functions related to the fuzzy controller. **a** Membership functions of fuzzy inputs. **b** Membership functions of fuzzy output

Table 5.2 Fuzzy logic rules regarding the angular velocity controller

F_l/F_r	Neg_{High}	Neg_{Low}	$Zero$	Pos_{Low}	Pos_{High}
Neg_{High}	$\omega_w = Z$	$\omega_w = Z$	$\omega_w = N_L$	$\omega_w = N_M$	$\omega_w = N_H$
Neg_{Low}	$\omega_w = Z$	$\omega_w = Z$	$\omega_w = N_L$	$\omega_w = N_L$	$\omega_w = N_M$
$Zero$	$\omega_w = P_L$	$\omega_w = P_L$	$\omega_w = Z$	$\omega_w = N_L$	$\omega_w = N_L$
Pos_{Low}	$\omega_w = P_M$	$\omega_w = P_L$	$\omega_w = P_L$	$\omega_w = Z$	$\omega_w = Z$
Pos_{High}	$\omega_w = P_H$	$\omega_w = P_M$	$\omega_w = P_L$	$\omega_w = Z$	$\omega_w = Z$

Seven functions were defined to the outputs as can be seen in Fig. 5.15b:

- $Negative_{High}(N_H)$, *Z-shaped* function with $a = -0.8$ and $b = 0.2$.
- $Negative_{Medium}(N_M)$, Gaussian symmetrical function with $\sigma = 0.2$ and $c = -0.4$.
- $Negative_{Low}(N_L)$, Gaussian symmetrical function with $\sigma = 0.2$ and $c = -0.2$.
- $Zero(Z)$, Gaussian symmetrical function with $\sigma = 0.2$ and $c = 0$.
- $Positive_{Low}(P_L)$, Gaussian symmetrical function with $\sigma = 0.2$ and $c = 0.2$.
- $Positive_{Medium}(P_M)$, Gaussian symmetrical function with $\sigma = 0.2$ and $c = 0.4$.
- $Positive_{High}(P_H)$, *S-shaped* function with $a = 0.8$ and $b = 0.2$.

A set of twenty-five rules were implemented in the fuzzy logic architecture as presented in Table 5.2.

After the fuzzy logic block, the signals are passed through the output conditioning block that performs two functions: (i) low pass filtering to avoid eventual abrupt changes in control signals and, thus, ensuring comfortable navigation to the user; and (ii) signal adjustments to obtain the walker's angular velocity range.

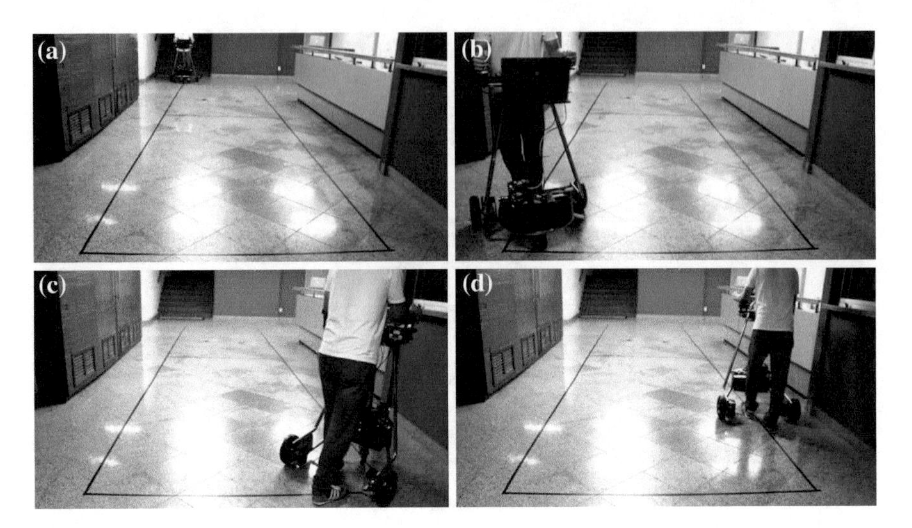

Fig. 5.16 Snapshot performing an u-shaped path by the user with the proposed control strategy

5.4.2 Controller Evaluation

An experiment was conducted with the proposed control strategy in order to evaluate its effectiveness. The user was asked to performed u-shaped path using the robotic walker with normal speed. Figure 5.16 shows snapshots of instants of such experiment. It is noteworthy that the u-shaped path is analyzed in four phases: first, a first straight path (Fig. 5.16a); second, a first curve turning to the left (Fig. 5.16b); third, a second curve turning to the left (Fig. 5.16c); Finally, a last straight path (Fig. 5.16d).

Figure 5.17 shows the control data recorded from the experiment during 50 s. In Fig. 5.17a, it is possible to observe lower values for the y-axis components when the subject is walking on a straight path (from 0 to 18th s and from 30th to 50th s). These intervals did not yield control actions on ω_w (Fig. 5.17b).

In Fig. 5.17a, the two left curves can be observed after the 18th and 25th s. A positive peak on F_{ly} and a negative peak on F_{ry} characterize such turning events. In the same manner, a control action to turn to the left is yield, which can be observed in ω_w (Fig. 5.17b). Finally, there is no significant delay between the control action, $\omega_w(C)$, and the measured angular velocity, $\omega_w(R)$.

To perform this experiment a desired distance dd equal to 0.5 m was selected, which was kept almost constant during all experiment (Fig. 5.17c) even when curves were performed: $\tilde{d}$ was always lower than 0.1 m.

In Fig. 5.17d, the control action (v_w) follows the human linear velocity (v_h), as expected. Finally, there is not a significant delay between the control action, $v_w(C)$, and the measured linear velocity, $v_w(R)$.

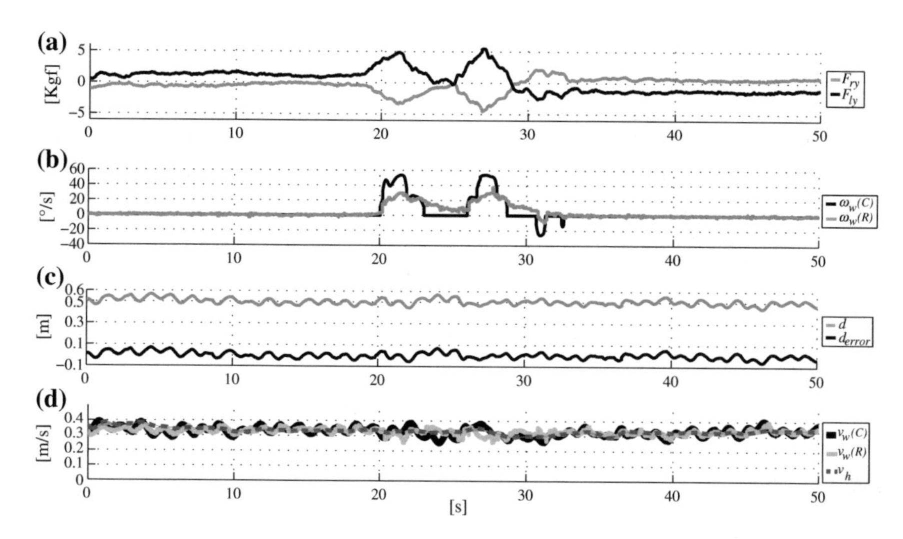

Fig. 5.17 Control data of a experiment conducted in an u-shaped path with the proposed control strategy. **a** F_{ly} and F_{ry}. **b** Angular velocities: control action $\omega_w(C)$ and measured $\omega_w(R)$. **c** Distances parameters. **d** Linear velocities: control action $v_w(C)$, measured $v_w(R)$ and v_h

5.5 Conclusions

This chapter presented the design and proof of concept of a multimodal interface that provides an online estimation of the human-walker interaction. The estimated parameters are used to drive a Smart Walker. Such multimodal sensor platform monitors the whole interaction through LRF, inertial sensor information, and 3D force sensors in order to attain a natural and reliable interface for the walker. The information provided by these different sensor technologies are fused according to a strategy aimed at providing a broad characterization of the walker-assisted gait phenomenon: legs location detection, human hip and walker orientation, and upper-limb interaction forces.

The proposed parameters show that it is possible to estimate the human velocity from the walker only using a LRF sensor, which was validated in the previous chapter. Additionally, combining signals obtained from the upper-limb guiding forces and the human hip orientation angle, it is also possible to monitor completely the motion chain when the user is turning from the passenger to locomotor units.

The parameters estimation was precise, showing also repeatability with continuous turns in the user path. In the same way, the mean rate of cadence rejection in the hip orientation and upper-limb guiding forces was 71 %.

The proposed filtering strategies and parameter estimation aims at developing more adaptable control strategies and safer robotic walker controllers. Such controllers will enable the development of functional compensation strategies in clinical environment. Furthermore, it constitutes a suitable framework to continuously monitor gait parameters for follow up of certain pathologies and assess the evolution of the rehabilitation processes.

This chapter also presented an example of a control strategy based on both pHRI and cHRI. Such controller utilizes force sensors and LRF (Laser Range Finder) to control a robotic walker without attaching any sensor on the user body. This approach combines user information about forearm reaction forces and gait kinematics from the legs scanning localization.

Remarkably, one of the main advantages of the proposed method is its computational efficiency. The estimated parameters do not present a considerable increase in the execution time. For this reason, this multimodal interface is suitable for real time control applications.

References

1. J. Perry, J. Burnfield, *Gait Analysis: Normal and Pathological Function*, 1st edn. (SLACK Incorporated, Grove Road, 1992). ISBN 978-1-55642-192-1
2. A. Frizera, R. Ceres, E. Rocon, J.L. Pons, A. Frizera-Neto, R. Ceres, E. Rocon, J.L. Pons, Empowering and assisting Natural human mobility: the simbiosis walker. Int. J. Adv. Robot. Syst. **8**(3), 34–50 (2011). http://oa.upm.es/13856/2/INVE_MEM_2011_115583.pdf
3. A. Abellanas, A. Frizera, R. Ceres, J.A. Gallego, Estimation of gait parameters by measuring upper limb-walker interaction forces. Sens. Actuators A: Phys. **162**(2), 276–283 (2010). ISSN 0924-4247
4. B. Widrow, S.D. Stearns, *Adaptive Signal Processing* (Prentice Hall, Englewood Cliffs, 1985)
5. A.F. Neto, J.A. Gallego, E. Rocon, Extraction of user's navigation commands from upper body force interaction in walker assisted gait. BioMed. Eng. OnLine, **9**(37) (2010). ISSN 1475-925X. doi:10.1186/1475-925X-9-37. http://www.pubmedcentral.nih.gov/articlerender.fcgi?artid=2924341&tool=pmcentrez&rendertype=abstract, http://www.biomedcentral.com/content/pdf/1475-925X-9-37.pdf
6. K. Ryoul Mun, H. Yu, C. Zhu, M.S.T.a Cruz, Design of a novel robotic over-ground walking device for gait rehabilitation, in *International Workshop on Advanced Motion Control* (AMC, 2014) pp. 458–463. ISBN 9781479923243. doi:10.1109/AMC.2014.6823325
7. Yu. Haoyong, Matthew Spenko, Steven Dubowsky, An adaptive shared control system for an intelligent mobility aid for the elderly. Auton. Robots **15**(1), 53–66 (2003)
8. M. Martins, C. Santos, A. Frizera, R. Ceres, Real time control of the ASBGo walker through a physical human-robot interface. Meas. J. Int. Meas. Confed. **48**(1), 77–86 (2014). ISSN 02632241. doi:10.1016/j.measurement.2013.10.031

Chapter 6
Conclusions and Future Works

As previously presented, there is a significant need to improve the ability of patients with gait impairments to promote safe and efficient ambulation. This work introduces some concepts that could be useful for the design of assistive and rehabilitation devices. Specifically, this book defines the concepts of physical and cognitive Human-Robot Interaction (HRI) for walker-assisted gait, with the aim of developing a more natural human-robot interaction.

Two new control strategies for HRI were proposed and validated. On the one hand, a control strategy for cognitive HRI during walking was presented using Laser Range Finder (LRF) and Inertial Measurement Units (IMU). A satisfactory result was obtained in terms of stable performance in the simulation environment. Such controller was implemented and validated in two robotic platforms: a mobile robot and a robotic walker. The controller keeps the robot continuously following in front of the human during walking in both implementations. Moreover, in the robotic walker evaluation, a comparison between the user guiding the walker and the walker following the user showed a similar behavior in terms of control errors. Consequently, this controller is suitable for natural human-robot interaction.

On the other hand, an implementation of a control strategy based on physical and cognitive HRI was presented. Such controller utilizes force sensors and a LRF to control a robotic walker without attaching any sensor on the user body. The controller keeps the walker continuously following in front of the user as a cognitive feature. Additionally, the physical interaction provides a more predictive behavior when the user performs curves, such as shown during the experimental validation.

Two methods for fusing LRF and IMU sensors to estimate the control inputs were proposed and validated. The first one is a human-robot interaction parameters detection synchronized with gait cycles was implemented for human tracking from a mobile robot. The second strategy relies on adaptive estimation and filtering of gait components. In the experimental studies, despite of the continuous body oscillation during walking, the parameters estimation was precise and unbiased. It also showed repeatability when speed changes and continuous turns were performed. Estimation errors were lower than 10 % in both methods.

C.A. Cifuentes and A. Frizera, *Human-Robot Interaction Strategies for Walker-Assisted Locomotion*, Springer Tracts in Advanced Robotics 115,
DOI 10.1007/978-3-319-34063-0_6

This work also presented the design and proof of concept of a multimodal interface that provides an online estimation of a human-walker interaction parameters. The estimated parameters are used to drive a Smart Walker. Such multimodal sensor platform monitors the whole interaction through LRF, inertial sensor information, and 3D force sensors in order to attain a natural and reliable interface for the walker. The information provided by these different sensor technologies are fused aiming at providing a broad characterization of the walker-assisted gait phenomenon, which includes legs location detection, human hip and walker orientation, and upper-limb interaction forces.

The parameters proposed in such multimodal interface show that is possible to estimate the human velocity from the walker only using a LRF sensor. Additionally, combining signals obtained from the upper-limb guiding forces and the human hip orientation angle, it is also possible to monitor completely the motion chain when the user is turning from the passenger to locomotor units.

The proposed filtering strategies and parameter estimation aim at developing more adaptable control strategies and safer robotic walker controllers. Such controllers will enable the development of functional compensation strategies in clinical environment. Furthermore, they constitute a suitable framework to continuously monitor gait parameters for follow up of certain pathologies and assess the evolution of the rehabilitation processes.

Remarkably, one of the main advantages of the proposed methods is its computational efficiency. The estimated parameters do not present a considerable increase in the execution time. For this reason, this multimodal interface is suitable for real time control applications.

As future work, a clinical protocol is being prepared to validate the interaction strategies and the robotic device with patients. Clinical evaluation and the adaptation of the interaction scheme is an important future task to clinical rehabilitation. Experiments will be conducted with people with motor disabilities to characterize the pathological walker-assisted gait and also to evaluate the interaction schemes developed.

Several refinements and extensions of the presented control strategies are conceivable. One potential improvement of the existing force interaction controller is the implementation of a variable admittance controller. Such as aforementioned in the Chap. 2, physical interaction can help in setting rules for cognitive evaluations of the environment during interaction tasks. For instance, a smart walker could provide the user different levels of force feedback according to different types of therapy, or regarding inadequate gait patterns.

As aforementioned in the Chap. 1, mobile gait rehabilitation devices that combine mobile platforms (smart walkers) with BWS system can enable free walking overground in different environments. That way, the cognitive HRI controller presented in this book has been implemented in a new robotic platform, which is a Wearable Robotic Walker (named CPWalker) to support novel therapies for Cerebral Palsy (CP) rehabilitation. This platform integrates a smart walker along with a passive

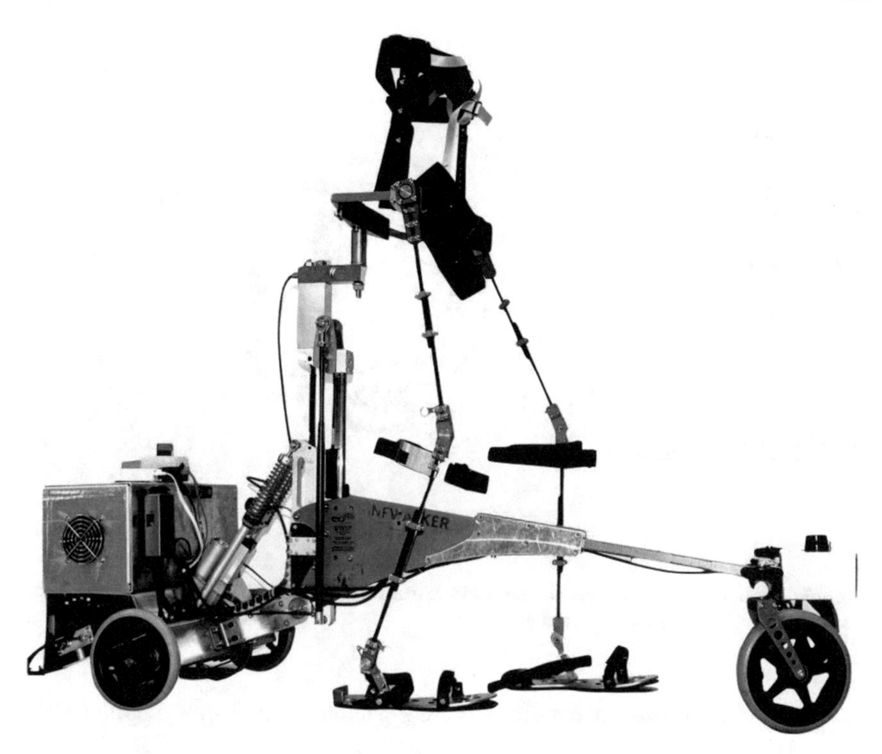

Fig. 6.1 CPWalker platform

lower-limb exoskeleton as a wearable device as can be seen in Fig. 6.1. CPWalker enables the use of a robotic platform through which the infant can start experiencing autonomous locomotion in a rehabilitation environment.

The robotic platform is based on the commercial available device named NF-Walker (Made For Movement, Norway), see Fig. 6.1. NFWalker is a passive Hands-free walker. The conceptual design drafted defined the mechanical modifications on the device NF-Walker. These actions were implemented in both the walker and the passive exoskeleton and resulted in the incorporation of the following systems: (i) a drive system of the platform; (ii) A discharge control system of the user weight; (iii) an active system for height control.

This approach aims at a natural and safe walker-assisted gait for training of CP patients. On the one hand, the robotic platform promotes an body weight support depending on both patient condition and therapy level. On the other hand, the control strategy intends to develop a natural human-robot interaction during the assisted locomotion. Although this platform is physically connected to the patient at the trunk, the user is able to perform a free leg motion with a proper gait pattern by means of a passive lower-limbs exoskeleton. Thus, the cognitive HRI controller uses the legs kinematics information to detect the users locomotion intentions to command the robots velocity. The platform is currently under validation with a group of CP patients at Niño Jesus Hospital in Madrid, Spain.

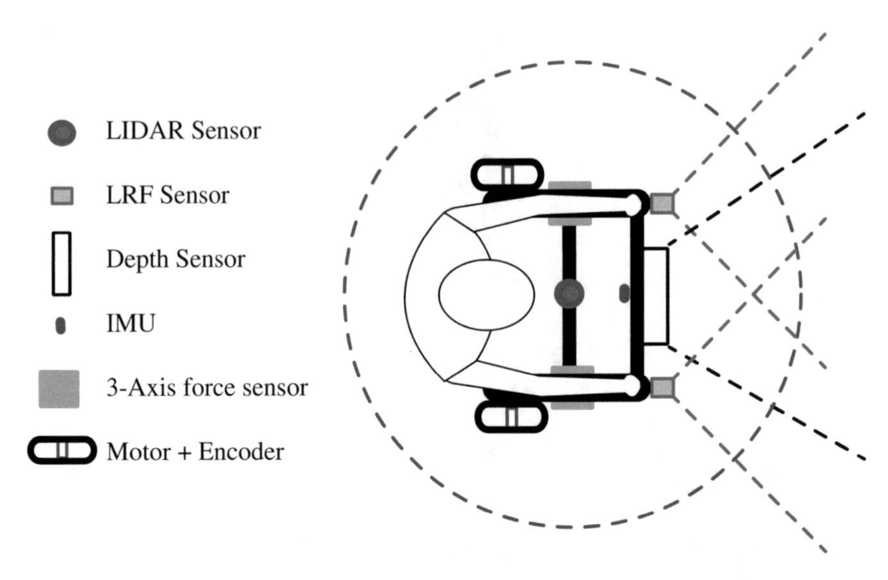

Fig. 6.2 Diagram that illustrates the human-environment interaction through a multimodal interface for social acceptability of smart walkers

As aforementioned, smart walkers present potential benefits for mobility assistance and gait rehabilitation, but it is also clear that the new generations of robots will work in close interactions with human beings. New robotic walkers should address the novel problem of social acceptability and intuitive human-robot interaction taking into account the environment. A proposal of a multimodal interface for human-environment interaction using a smart walker is presented in Fig. 6.2, which is based on mobile robot approaches for SLAM (Simultaneous Localization and Mapping) strategies and navigation. This interface combines different sensors, such as LRF, IMU, LIDAR (Light Detection and Ranging) and depth sensors.

However, new technological breakthroughs are also required, such as: (i) control strategies for adapting to dynamic and open environments populated by human beings and (ii) the sensor and control levels should deal with incompleteness and uncertainty. In fact, real world situations are highly complex for being fully modeled using classical tools (e.g., kinematics and dynamics approaches). Consequently, it is necessary to introduce probabilistic reasoning approaches in the control architecture, which is an emerging topic of research in the human-robot interaction field.

Additionally, the integration of active orthoses or robotic exoskeletons has been considered in this approach. These devices provide external controlled power to the impaired joints to compensate walking function. This way, the robotic exoskeleton may be used to substitute or enhance motor function of the lower limbs for walking by driving the user's joints through a functional walking pattern. The combination of a lower-limb exoskeleton with a smart walker is showed in Fig. 6.3. Thus, the smart walker allows the partial body weight support during the walk, and the robotic exoskeleton performs the gait control with a physiological gait pattern.

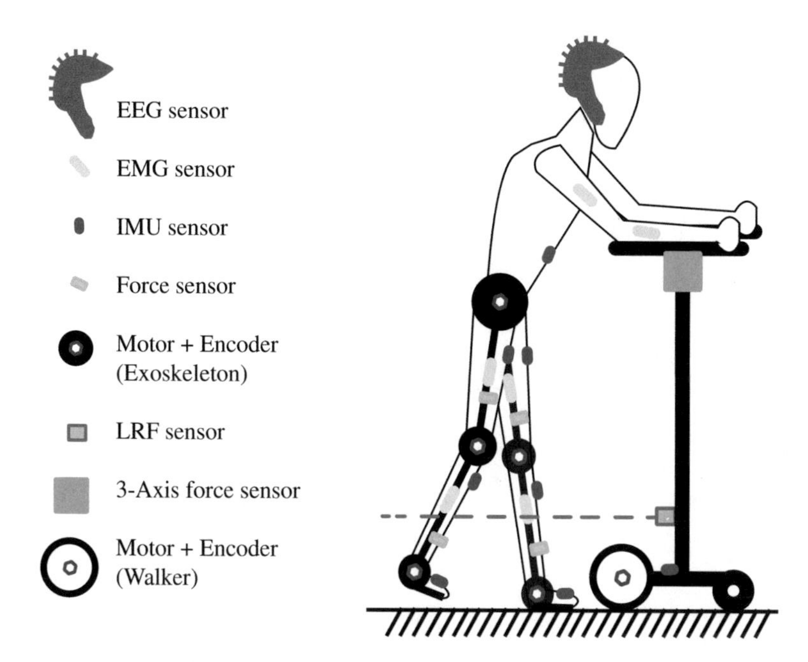

Fig. 6.3 Diagram that illustrates the combination of a multimodal interface, a lower-limb exoskeleton and a smart walker

In the Chap. 5, a comparison between hip motion information and upper-limb reaction forces shows that the interaction forces contain information more predictable regarding the human motor intentions. However, the integration of other sensory modalities related to human motor control, including neural signals from the central nervous system (e.g., electroencephalography), neural muscular activation (EMG) and other sensor interfaces could enable more predictable control strategies as illustrated in Fig. 6.3.

Printed in the United States
By Bookmasters